Energy Storage Technologies in Grid Modernization

Scrivener Publishing
100 Cummings Center, Suite 541J
Beverly, MA 01915-6106

Publishers at Scrivener
Martin Scrivener (martin@scrivenerpublishing.com)
Phillip Carmical (pcarmical@scrivenerpublishing.com)

Energy Storage Technologies in Grid Modernization

Edited by
Sandeep Dhundhara
Yajvender Pal Verma
and
Ashwani Kumar

WILEY

This edition first published 2023 by John Wiley & Sons, Inc., 111 River Street, Hoboken, NJ 07030, USA and Scrivener Publishing LLC, 100 Cummings Center, Suite 541J, Beverly, MA 01915, USA
© 2023 Scrivener Publishing LLC
For more information about Scrivener publications please visit www.scrivenerpublishing.com.

All rights reserved. No part of this publication may be reproduced, stored in a retrieval system, or transmitted, in any form or by any means, electronic, mechanical, photocopying, recording, or otherwise, except as permitted by law. Advice on how to obtain permission to reuse material from this title is available at http://www.wiley.com/go/permissions.

Wiley Global Headquarters
111 River Street, Hoboken, NJ 07030, USA

For details of our global editorial offices, customer services, and more information about Wiley products visit us at www.wiley.com.

Limit of Liability/Disclaimer of Warranty
While the publisher and authors have used their best efforts in preparing this work, they make no representations or warranties with respect to the accuracy or completeness of the contents of this work and specifically disclaim all warranties, including without limitation any implied warranties of merchantability or fitness for a particular purpose. No warranty may be created or extended by sales representatives, written sales materials, or promotional statements for this work. The fact that an organization, website, or product is referred to in this work as a citation and/or potential source of further information does not mean that the publisher and authors endorse the information or services the organization, website, or product may provide or recommendations it may make. This work is sold with the understanding that the publisher is not engaged in rendering professional services. The advice and strategies contained herein may not be suitable for your situation. You should consult with a specialist where appropriate. Neither the publisher nor authors shall be liable for any loss of profit or any other commercial damages, including but not limited to special, incidental, consequential, or other damages. Further, readers should be aware that websites listed in this work may have changed or disappeared between when this work was written and when it is read.

Library of Congress Cataloging-in-Publication Data

ISBN 9781119872115

Front cover images supplied by Pixabay.com
Cover design by Russell Richardson

Set in size of 11pt and Minion Pro by Manila Typesetting Company, Makati, Philippines

Printed in the USA

10 9 8 7 6 5 4 3 2 1

Contents

Preface		xiii
1	**Overview of Current Development and Research Trends in Energy Storage Technologies**	**1**
	O. Apata	
	1.1 Introduction	1
	1.2 The Technology of Energy Storage	4
	1.3 Energy Storage and Smart Grids	14
	1.4 Energy Storage and Micro-Grids	15
	1.5 Energy Storage Policy Recommendations	17
	1.6 Energy Storage: Challenges and Opportunities	18
	1.7 Practical Implementations of Energy Storage Technologies	19
	1.8 Conclusions	20
	References	20
2	**A Comprehensive Review of the Li-Ion Batteries Fast-Charging Protocols**	**23**
	Talal Mouais and Saeed Mian Qaisar	
	2.1 Introduction	24
	2.2 The Literature Review	27
	2.2.1 Overview of Lithium-Ion Battery Working Principle	28
	2.2.2 Principles of Battery Fast-Charging	31
	2.2.3 Multi-Scale Design for Fast Charging	33
	2.2.4 Electrode Materials	33
	2.2.5 Fast-Charging Strategies	34
	2.2.6 Types of Charging Protocols	34
	2.2.7 Li-Ion Battery Degradation	40
	2.2.8 Factors that Cause Battery Degradation	41
	2.2.9 Degradation Mechanism of the Li-Ion Battery	44
	2.2.10 Electrode Degradation in Lithium-Ion Batteries	48
	2.2.11 The Battery Management System	50

		2.2.12	Battery Technology Gap Assessment for Fast-Charging	53
		2.2.13	Developmental Needs	55
	2.3	Materials and Methods		56
	2.4	Discussion		58
	2.5	Conclusion		63
		Acknowledgements		65
		References		65

3 Development of Sustainable High-Performance Supercapacitor Electrodes from Biochar-Based Material   71
Kriti Shrivastava and Ankur Jain

3.1	Introduction		72
3.2	Role of Energy Storage Systems in Grid Modernization		73
3.3	Overview of Current Developments of Supercapacitor-Based Electrical Energy Storage Technologies		78
3.4	Potential of Biochar as High-Performance Sustainable Material		80
3.5	Overview of Recent Developments in Biochar-Based EDLC Supercapacitor		83
	3.5.1	Wood & Plant Residues as Biochar Precursor for Supercapacitor Applications	84
	3.5.2	Biochar-Based Supercapacitors from Waste Biomass	89
	3.5.3	Carbon-Based Supercapacitors from Other Methods	91
3.6	Current Challenges and Future Potential of Biochar-Based Supercapacitor		93
3.7	Conclusion		99
	References		101

4 Energy Storage Units for Frequency Management in Nuclear Generators-Based Power System   105
Boopathi D., Jagatheesan K., Sourav Samanta, Anand B. and Satheeshkumar R.

4.1	Introduction		105
	4.1.1	Structure of the Chapter	110
	4.1.2	Objective of the Chapter	110
4.2	Investigated System Modeling		111
	4.2.1	Battery Energy Storage System (BESS) Model	112
	4.2.2	Fuel Cell (FC) Model	113
	4.2.3	Redox Flow Battery (RFB) Model	113

		4.2.4	Proton Exchange Membrane (PEM) Based FC Model	114
		4.2.5	Ultra-Capacitor (UC) Model	115
		4.2.6	Supercapacitor Energy Storage (SCES) Model	116
	4.3	Controller and Cost Function		116
	4.4	Optimization Methodology		118
	4.5	Impact Analysis of Energy Storage Units		119
		4.5.1	Impact of BESS	119
		4.5.2	Impact of FC	121
		4.5.3	Impact of RFB	122
		4.5.4	Impact Analysis of the PEM-FC	123
		4.5.5	Impact Analysis of UC	125
		4.5.6	Impact Analysis of SCES	127
	4.6	Result and Discussion		128
	4.7	Conclusion		130
		Appendix		132
		References		132

5 Detailed Comparative Analysis and Performance of Fuel Cells 135

Tejinder Singh Saggu and Arvind Dhingra

	5.1	Introduction		135
	5.2	Classification of Fuel Cells		136
		5.2.1	Based on Fuel-Oxidizer Electrolyte	138
			5.2.1.1 Direct Fuel Cell	138
			5.2.1.2 Regenerative FC	139
			5.2.1.3 Indirect Fuel Cells	143
		5.2.2	Based on the State of Aggregation of Reactants	144
			5.2.2.1 Solid Fuel Cells	144
			5.2.2.2 Gaseous Fuel Cells	145
			5.2.2.3 Liquid Fuel Cells	147
		5.2.3	Based on Electrolyte Temperature	148
			5.2.3.1 Proton Exchange Membrane	148
			5.2.3.2 Direct Methanol	150
			5.2.3.3 Alkaline	150
			5.2.3.4 Phosphoric Acid	151
			5.2.3.5 Molten Carbonate	152
			5.2.3.6 Solid Oxide	153
	5.3	Cost of Different Fuel Cell Technologies		154
	5.4	Conclusion		155
		References		155

6 Machine Learning–Based SoC Estimation: A Recent Advancement in Battery Energy Storage System   **159**
Prerana Mohapatra, Venkata Ramana Naik N. and Anup Kumar Panda

- 6.1 Introduction   160
- 6.2 SoC Estimation Techniques   163
  - 6.2.1 Coulomb Counting Approach   164
  - 6.2.2 Look-Up Table Method   164
  - 6.2.3 Model-Based Methods   164
    - 6.2.3.1 Electrochemical Model   164
    - 6.2.3.2 Equivalent Circuit Model   165
  - 6.2.4 Data-Driven Methods   165
  - 6.2.5 Machine Learning–Based Methods   166
    - 6.2.5.1 Support Vector Regression   166
    - 6.2.5.2 Ridged Extreme Learning Machine (RELM)   168
- 6.3 BESS Description   171
- 6.4 Results and Discussion   171
- 6.5 Conclusion   175
- References   177

7 Dual-Energy Storage System for Optimal Operation of Grid-Connected Microgrid System   **181**
Deepak Kumar and Sandeep Dhundhara

- 7.1 Introduction   182
- 7.2 System Mathematical Modelling   188
  - 7.2.1 Modelling of Wind Turbine Power Generator   189
  - 7.2.2 Modelling of Solar Power Plant   189
  - 7.2.3 Modelling of Conventional Diesel Power Generator   189
  - 7.2.4 Modelling of Combined Heat and Power (CHP) and Boiler Plant   190
  - 7.2.5 Modelling of Dual Energy Storage System   190
    - 7.2.5.1 Battery Bank Storage System   190
    - 7.2.5.2 Pump Hydro Storage System   191
  - 7.2.6 Modelling of Power Transfer Capability   191
- 7.3 Objective Function and Problem Formulations   192
  - 7.3.1 Operational and Technical Constraints   192
- 7.4 Simulation Results and Discussion   195
- 7.5 Conclusion   208
- References   209

8 Applications of Energy Storage in Modern Power System through Demand-Side Management — 213
Preeti Gupta and Yajvender Pal Verma

- 8.1 Introduction to Demand-Side Management — 214
  - 8.1.1 Demand-Side Management Techniques — 214
    - 8.1.1.1 Energy Efficiency — 214
    - 8.1.1.2 Demand Response — 215
  - 8.1.2 Demand-Side Management Approaches — 217
- 8.2 Operational Aspects of DR — 218
- 8.3 DSM Challenges — 221
- 8.4 Demand Response Resources — 223
- 8.5 Role of Battery Energy Storage in DSM — 224
  - 8.5.1 Case Study I: Peak Load and PAR Reduction — 225
    - 8.5.1.1 Problem Formulation — 225
    - 8.5.1.2 Energy Storage Dispatch Modelling — 226
  - 8.5.2 Case Study II: Minimizing Load Profile Variations — 229
    - 8.5.2.1 Problem Formulation — 229
    - 8.5.2.2 SPV System Modelling — 230
  - 8.5.3 Results and Discussions — 231
    - 8.5.3.1 Case Study I: Peak Load and PAR Reduction Using Batteries with DR — 231
    - 8.5.3.2 Case Study II: Minimizing Load Profile Variations Using Batteries with DR — 232
- 8.6 Conclusion — 234
- References — 234

9 Impact of Battery Energy Storage Systems and Demand Response Program on Locational Marginal Prices in Distribution System — 239
Saikrishna Varikunta and Ashwani Kumar

- 9.1 Introduction — 240
  - 9.1.1 Battery Energy Storage System (BESS) — 240
  - 9.1.2 Demand Response Program — 242
- 9.2 Problem Formulation and Solution Using GAMS — 244
  - 9.2.1 Objective Functions for Case Studies: Case 1 to Case 5 — 245
    - 9.2.1.1 Case 1: Is Minimization of the Active Power Production Cost — 245
    - 9.2.1.2 Case 2: Minimization of the Active Power Production and Reactive Power Production Cost — 246

        9.2.1.3   Case 3: Minimization of the Active
                  Power Production and Reactive Power
                  Production Cost Along
                  with Capacitor Placement   246
        9.2.1.4   Case 4: Minimization of the Active Power
                  Production and Reactive Power Production
                  Cost Including Capacitor and BESS Cost   247
        9.2.1.5   Case 5: Minimization of the Active Power
                  Production and Reactive Power Production
                  Cost Including Capacitor and BESS Cost
                  and Taking the Impact of Demand
                  Response Program   248
    9.2.2 Real and Reactive Power Equality Constraints   249
        9.2.2.1   Equality Constraints   249
        9.2.2.2   Inequality Constraints: (at any bus i):
                  Voltage, Power Generation, Line Flow,
                  SOC, Battery Energy Storage Power   250
    9.2.3 Modified Lagrangian Function   251
    9.2.4 Generator Economics Calculations   252
  9.3 Case Study: Numerical Computation   254
  9.4 Results and Discussions   257
    9.4.1 Case 1: Minimization of the Active Power
           Production Cost   257
    9.4.2 Case 2: Minimization of the Active Power
           Production and Reactive Power Production Cost   260
    9.4.3 Case 3: Minimization of the Active Power
           Production and Reactive Power Production
           Cost Along   262
    9.4.4 Case 4: Minimization of the Active Power
           Production and Reactive Power Production Cost   266
    9.4.5 Case 5: Minimization of the Active Power
           Production and Reactive Power Production Cost   269
  9.5 Conclusions   279
       References   280

10 **Cost-Benefit Analysis with Optimal DG Allocation
   and Energy Storage System Incorporating Demand
   Response Technique**   283
   *Rohit Kandpal, Ashwani Kumar, Sandeep Dhundhara
   and Yajvender Pal Verma*
   10.1 Introduction   284

10.2	Distribution Generation and Energy Storage System		285
	10.2.1	Renewable Energy in India	286
	10.2.2	Different Types of Energy Storage and their Opportunities	287
	10.2.3	Distributed Generation	290
		10.2.3.1 Solar Photovoltaic Panel-Based DG (PVDG)	290
		10.2.3.2 Wind Turbine–Based DG (WTDG)	291
		10.2.3.3 Load Model and Load Profile	293
	10.2.4	Demand Response Program	294
	10.2.5	Electric Vehicles	297
	10.2.6	Modeling of Energy Storage System	299
	10.2.7	Problem Formulation	300
	10.2.8	Distribution Location Marginal Pricing	301
10.3	Grey Wolf Optimization		302
10.4	Numerical Simulation and Results		304
10.5	Conclusions		312
	References		313

11 Energy Storage Systems and Charging Stations Mechanism for Electric Vehicles — 317
Saurabh Ratra, Kanwardeep Singh and Derminder Singh

11.1	Introduction to Electric Vehicles		318
	11.1.1	Role of Electric Vehicles in Modern Power System	318
	11.1.2	Various Storage Technologies	319
	11.1.3	Electric Vehicle Charging Structure	322
11.2	Introduction to Electric Vehicle Charging Station		323
	11.2.1	Types of Charging Station	323
	11.2.2	Charging Levels	324
	11.2.3	EV Charging	324
	11.2.4	Charging Period	327
11.3	Modern System Efficient Approches		328
	11.3.1	Smart Grid Technology	328
	11.3.2	Renewable Energy Technology	329
	11.3.3	V2G Technology	329
	11.3.4	Smart Transport System	329
11.4	Battery Charging Techniques		330
	11.4.1	Electric Vehicle Charging Station in Modern Power System	331
11.5	Indian Scenario		332
11.6	Energy Storage System Evaluation for EV Applications		333

11.7	ESS Concerns and Experiments in EV Solicitations		334
	11.7.1	Raw Materials	335
	11.7.2	Interfacing by Power Electronics	335
	11.7.3	Energy Management	335
	11.7.4	Environmental Impact	336
	11.7.5	Safety	336
11.8	Conclusion		336
	References		337

Index   **341**

Preface

The electrical infrastructure of the grid is aging and there is a need to modernize the existing grid meeting the challenges of the 21st century power sector operation with the penetration of renewable energy sources. Modernization of the grid by incorporating smart grid technologies and moving towards grid digitization will make the power grid resilient to avoid power outages under extreme weather conditions and climatic changes. Energy storage technologies have emerged as a promising solution to address the challenges of grid modernization. These technologies have gained significant attention in recent years due to their potential to enable the integration of renewable energy sources and enhance the resilience, reliability, and flexibility of the electric grid. With the increasing adoption of intermittent renewable energy sources, the need for energy storage technologies has become more critical, as they can help mitigate the variability and uncertainty of these sources and enable their integration into the grid. These technologies enable the efficient and reliable storage of excess energy during times of low demand and the discharge of that energy during times of high demand. Energy storage technologies will act as an ancillary services to the grid for frequency regulation, voltage support, and providing black start capabilities.

This book provides a comprehensive overview of the various energy storage technologies, their applications in grid modernization, and the simulation-based case studies showing their implementation in modern power systems. It covers the overview of current development and research trends and various applications of energy storage technologies and their role in enabling a more sustainable and resilient energy system. The book contains eleven chapters that cover the specific aspect of energy storage technologies in grid modernization. The chapters provide a clear and concise explanation of the key concepts, illustrated by real-world examples and case studies.

In chapter 1, an overview of energy storage technologies, recent trends, role of energy storage in smart/micro grids is presented. The chapter provides opportunities and challenges in energy storage, policy recommendations and practical applications of ESS. Chapter 2 provides comprehensive overview of Li-ion batteries, its materials, operation, charging protocols and fast charging strategies. Chapter 3 deals with an overview of current developments of Super-capacitor-based electrical energy storage technologies. Recent Developments in Biochar-Based EDLC Supercapacitor, Carbon-Based Supercapacitors and current challenges and future potential is discussed.

Chapter 4, deals with the role of energy storage technology in frequency regaultion services. The impact on frequency regulation due to battery energy storage systems, fuel cell, Supercapacitor Energy Storage, and ultra capacitors are analyzed. Chapter 5 provides detailed comparative analysis and performance of Fuel Cells. Chapter 6 described the Machine Learning–Based SoC estimation and recent advances in Battery Energy Storage System techniology. Electrical equivalent and electrochemical model of the BESS are discussed and Machine Learning–Based methods as Support Vector Regression (SVR) and Ridged Extreme Learning Machine (RELM) are applied to SoC estimation. In the chapter 7, dual-energy storage system for optimal operation of grid-connected Microgrid System is analyzed. The modeling of different renewable energy sources are described and role of storage technology on power transfer capability is presented. Chapter 8 deals with the applications of energy storage in modern power system using demand-side management for peak load management, minimizing load variations, and energy storage dispatching. The case studies are provided to show the impact of energy storage technology on the load management. Chapter 9 presents impact of battery energy storage and demand response program on locational marginal prices in distribution system. Five different case studies have been analyzed to show the impact of energy storage systems on the active and reactive power cost in a distribution system. Chapter 10 gives cost-benefit analysis with optimal DG allocation and energy storage system considering the role of demand response technique. The impact of distributed generation from solar PV and wind power generators on system operation and locational marginal prices have been analyzed considering the electrical vehicles in the network. Chapter 11 discusses energy storage systems and charging stations mechanism for Electric Vehicles. The role of energy storage system and its evaluation for EV applications are presented in the context of Indian scenario.

This book is intended for a broad audience, including researchers, engineers, policymakers, and industry professionals working in the field of energy storage technologies and grid modernization. It provides a comprehensive and up-to-date reference for understanding the current state of the art, the challenges and opportunities, key role of energy storage technology in operation and management of smart/micro grids. It provides the potential platform for an innovation and deployment of energy storage technologies in the electric power grid.

We are extremely thankful to all the authors who have contributed their chapters to this book. We sincerely thank all excellent reviewers to give their valuable time providing their crtical evaluation, suggestions and comments for the improvement of the chapters. We sincerely acknowledge their support in the review process and bringing the book to the final stage.

Sandeep Dhundhara
Yajvender Pal Verma
Ashwani Kumar

1

Overview of Current Development and Research Trends in Energy Storage Technologies

O. Apata

Department of Electrical Engineering, Independent Institute of Education, IIEMSA, Roodepoort, South Africa

Abstract

The role of energy storage in ensuring grid flexibility and security of energy supply cannot be over-emphasized. Energy storage technologies harvest the available intermittent power from renewable energy sources in times of excess to be redistributed during scarcity by decoupling energy supply and demand, therefore improving grid flexibility, resiliency, and reliability. Different types of energy storage technologies have been proposed for grid integration of renewable energy sources. This chapter presents an overview of the various storage technologies, providing a comparative analysis of the different energy technologies and their application to smart grids, paying attention to the pros and cons of each of these technologies. This chapter also presents discussions around current developments and trends in this ever-evolving research area. It is important to state that environmental benefits can be provided indirectly by energy storage technologies. It is therefore pertinent to also understand the net environmental impact of the different energy storage technologies.

Keywords: Energy storage, renewable energy sources, smart grid, micro-grid

1.1 Introduction

Over the last decade, there has been a rapid growth in population and urbanization which has led to a corresponding exponential demand for

---

Email: g.apata@ieee.org

Sandeep Dhundhara, Yajvender Pal Verma, and Ashwani Kumar (eds.) *Energy Storage Technologies in Grid Modernization*, (1–22) © 2023 Scrivener Publishing LLC

energy. The exponential demand for energy across the globe has also raised concerns about greenhouse gas (GHG) emissions and their impact on climate change. To address the rising concern of climate change and greenhouse gas emissions, the United Nations through its climate action plan came up with a strategy in 2015, known as the Paris Agreement [1, 2]. This action plan seeks to substantially reduce GHG, especially from fossil fuels. Governments all over the world are therefore embracing renewable energy solutions as alternatives to conventional fossil fuel plants to meet the rising energy demand. In its 2021 report, the International Energy Agency (IEA), indicated global installations of new renewable energy power increased to 290 GW compared to 280 GW in 2020, representing a 3% growth in renewable energy installations globally [3]. Figure 1.1 below shows the growth in the use of renewable energy sources in the last eight years.

However, a significant concern of renewable energy sources, wind and solar, for energy generation is the intermittency of these renewable energy sources and power fluctuations, increasing the complexities associated with planning and operation of the grid. This has been the focal point of argument for those who are against a transition to alternative/renewable energy sources. However, the concerns of energy intermittency from renewable energy sources can be solved by different solutions such as load shifting by demand management, electric energy storage, interconnection with external grids, and a few others. Energy storage systems (ESS) have been identified as a promising approach to the challenges associated with different renewable energy sources. ESS can help in the mitigation of power variations, enabling the storage and dispatch of electrical energy

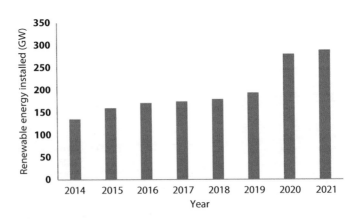

Figure 1.1 Global installed renewable energy capacity.

generated from renewable energy sources and ultimately improving the system flexibility.

Electric energy storage entails converting electrical energy into a storable form, and reserving it in different mediums, converting the stored energy back into electrical energy when needed. The integration of ESS into renewable energy systems not only helps to alleviate the intermittency of the renewable source but also reduces energy import during periods of peak demand and provides time-varying energy management. The concept of energy storage for bulk power supply has been in existence as far back as the 1930s when batteries were used in stabilizing and providing support to the power system in different German cities, especially at night [4]. The Electric storage batteries at that time provided the energy needs and emergency capacity during the peak periods when the dynamos, which generated direct current, were out of service.

The International Energy Agency (IEA) in its 2021 report [5] projected a 56% expansion in the global installed energy storage capacity in the next five years, reaching over 270 GW in 2026 with utility-scale batteries projected to experience significant growth in energy storage capacity worldwide. In the study carried out in [6], the report projected that by the end of 2030, the total cumulative installed energy storage capacity would be 741 GWh representing an exponential 31% compound annual growth rate (CAGR) of globally installed energy storage capacity. This is a pointer to the importance of EES to the modern-day power system. The rapid scaling up of energy storage systems is therefore of great importance to address the variability of renewable energy sources as their share of generation increases on the road to a net-zero carbon emission by 2050.

The different services which can be obtained from the integration of ESS include but are not limited to customer energy management services, ancillary services, bulk energy services, transmission infrastructure services and off-grid applications. The integration of EES in both isolated and grid-connected scenarios has multiple advantages ranging from the improvement of power quality to off-grid services such as electric vehicles (EV) and micro-grid stability. With ESS, an opportunity is presented for peak load demand shaving and demand-side management (DSM) ultimately reducing the burden associated with the installation of new generation capacity. EVs are mobile EES and complement the storage capacity of the grid while plug-in-electric (PEV) and smart electric parks have the potential to reduce the peak-to-average ratio and the overall electricity cost in real-time pricing by participation in energy trading of DSM.

The rest of this chapter discusses the different aspects of electric energy storage.

1.2 The Technology of Energy Storage

An ideal energy storage technology should be cost-effective and have a short start-up time, have high roundtrip efficiency, be able to operate at the level of the electric grid with a minimal power rating, and must have an appropriate capacity over power ratio for load shifting. It is important to note that none of the currently existing technologies for ESS has all of the above-mentioned characteristics at once, therefore a trade-off is required depending on need. The choice of storage technology adopted is dependent on factors such as storage duration, end application, the type of energy production, charging and discharging rate, and the depth of the chosen system [7].

Figure 1.2 presents a classification of the different energy storage technologies based on energy usage in a specific form. Each system has its distinguishable characteristic in wattage rating, life cycle, energy density, discharge time, and discharge loss. These characteristics are important to determine the suitability of each storage technology in providing the different services described in Section 1.1.

A. Mechanical energy storage
This storage technology is advantageous because of the flexibility it provides in converting between mechanical and electrical energy forms.

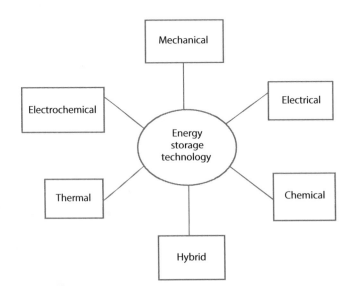

Figure 1.2 Classification of energy storage technologies.

Using the principle of potential and kinetic energy, and in some instances, pressurized gas, during off-peak periods electric energy is converted into mechanical energy when energy demand is low and converted back into electrical energy during peak demand for electrical energy. The most commonly available mechanical energy storage are: gravity energy storage systems, compressed air energy storage, flywheel energy storage system, and pumped hydro storage.

1. Compressed air energy storage (CAES)
The CAES operates on the principle of operation of gas turbine systems. In this energy storage system, energy is stored either as mechanical energy or a combination of thermal and mechanical energy. By compressing air and storing it in an underground space such as a cavern underground, the storage of energy is achieved. Electricity is produced when there is an expansion of the modified gas to rotate the turbine. During off-peak power demand, by utilizing the available excess power, a generator unit or reversible motor

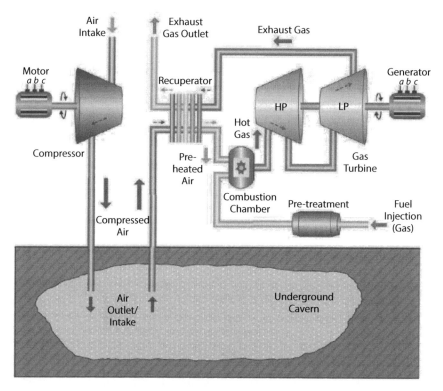

Figure 1.3 Schematic of a large-scale diabatic in-ground cavern CAES system [10].

is driven. This in turn injects air into the storage unit by running a chain of compressors. The compressed air is released and heated by a heat source during low power generation for load demand before being transferred to the turbine where a recuperation unit recycles the waste heat energy.

CAES systems are ideal for small- to large-scale power capacity; however, they can be deployed for large-scale applications involving peak shaving, voltage control, frequency control, and load shifting [8]. CAES can be further classified into diabatic, adiabatic, and isothermal storage [9] based on endothermic and exothermic processes that take place in the compression and expansion of air and heat exchange. System flexibility and high power density are two important characteristics of the Diabatic CAES system that make them the most commercially implemented CAES system. The CAES system can maintain the required system temperature by using external power sources in the heating and cooling of the air as shown in Figure 1.3; this makes them highly efficient. Adiabatic and isothermal CAES systems are well positioned for systems requiring small power density.

CAES systems are highly efficient thermodynamic systems due to the continuous subtraction and addition of heat during compression and expansion, respectively, keeping the air at ambient values. In addition to the ability to smoothen the power output of renewable energy plants, CAES has a high response time. However, a major challenge associated with CAES systems is the selection of a suitable geographical location for such a project.

2. Gravity energy storage (GES) systems
This is an emerging field in energy storage systems that has become a popular alternative to CAES and pumped hydro systems for large-scale power systems. The basic structure of a GES is represented in Figure 1.4. This is a closed system consisting of a piston, generator, reversible turbine/pump, and a container with a returned pipe.

When energy demand is high, water is pushed to flow into the container by the piston to drive the pump/turbine. The kinetic energy of water is then converted into mechanical energy by the turbine, spinning the generator to drive the turbine to produce electricity. A reverse mechanism is initiated during off-peak demand by supplying the excess energy to the motor. The piston is driven to the top of the container by the generated kinetic energy, and the mechanical energy is stored. The GES storage system overcomes the geographical limitation associated with the CAES system.

3. Pumped hydro storage (PHS)
The principle of operation of the PHS is based on storing electrical energy as a form of potential energy by pumping water from the lower side of a

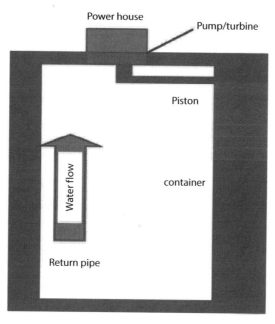

Figure 1.4 Basic structure of a GES [11].

reservoir to its higher side when energy demand is lower and vice versa during high energy demand. The performance of a PHS system is generally dependent on the volume and height of available water. Energy storage using PHS is achieved by pumping water uphill using off-peak electricity and then allowing the water to flow downhill, driving the generator to produce electricity for the power grid when the need arises. It is mostly applied in storing and generating electricity using two water reservoirs at different elevations, recompensing high-peak demand. PHS is a commercially available technology representing about 99% of installed EES capacity.

A typical PHS system consists of a generator, upper reservoir, lower reservoir, an inlet valve, penstock valve, motor, and pump. Energy storage time in the PHS is prolonged since the application process is subdivided into a 24-hour scale. The power rating of a PHS varies between 1 MW to 3000 MW at an operating efficiency of 76-85% with practically unlimited life cycles and an operating life span of about 50 years. PHS can be subdivided into three major classifications:

(a) Open PHS system: This is a pump back system that allows water to continuously flow through the upper and lower reservoir.

(b) Semi-open PHS: This is made up of one modified lake with continuous through flow and one modified or artificial reservoir.
(c) Closed loop PHS: This system comprises two non-connected reservoirs and is split by a vertical span.

A significant advantage of the PHS is its remarkable fast response time, which is typically less than a minute, enabling the PHS system to become an important component controlling electrical network frequency in the provision of reserve generation. PHS can act as a stabilizer for power systems consisting of renewable energy sources through its flexible control. The main drawback of the PHS is the constraint of geographical availability.

4. Flywheel energy storage (FES) system

This energy storage system is made of a large cylinder fixed on a stator by magnetic glide bearings that serve as the main component of the FES systems as shown in Figure 1.5. The storage of energy in the FES is a mechanically executed process because the kinetic energy of the rotor mass spins at very high speeds. Through torque control, the energy stored in the flywheel can be reused by reducing the speed of the flywheel while the kinetic energy returns to the electrical motor, functioning as an electric generator. The energy produced by the FES is dependent on the moment of inertia of the rotor and the speed at which it is rotated along its tensile strength and stress restrictions.

FES can be generally classified into low- and high-speed FES systems. Low-speed FES are generally used in power quality applications requiring high power for short durations with a high number of charge-discharge

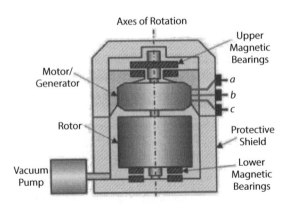

Figure 1.5 Components of a flywheel storage system [12].

cycles. The high-speed FES system on the other hand has a high energy density in the range of 200 Wh/kg with a high-power density. In contrast to the low-speed FES, which is cheap and commercialized, the high-speed FES is not economical because of the high cost associated with high-speed composite materials, making their use limited to specific longer storage systems.

Some merits of the FES include a performance rate of about 90% and a longer cycle life (some FES have a capability of over 100 000 full discharge cycle depth), operating at varying temperature conditions with the ability to sustain high power levels. However, a major drawback of the FES is the "flywheel explosion". This is a scenario where the flywheel tensile strength is exceeded causing the flywheel to shatter, and consequently releasing all of its stored energy at one go.

B. Electrical energy storage (EES) systems

These storage systems can store electrical energy for the production of electric energy and supply the same to the load for use when required. By modifying the magnetic or electric fields using superconducting magnets or capacitors, energy is stored. Since power systems face numerous challenges in the integration of renewable energy sources into the transmission and distribution systems, EES systems have been proposed as an appropriate technology to mitigate these challenges due to its different operating features such as the ability to support micro-grids, reduction of electrical energy import during peak demand period, load balancing and improving power quality. An ultra-capacitor also referred to as a super-capacitor and the super magnetic energy storage are typical examples of an EES.

An ultra-capacitor (UC) stores electrical energy between two conducting electrodes, as shown in Figure 1.6. An advantage of this technology is the absence of any chemical reactions, making it an alternative to the typical capacitor used in general batteries and different electronic applications. The UC operates with a large surface area and molecule thin layer of electrolyte. The UC also has a high peak power output with a high-power density. In comparison to a conventional battery, the UC has a longer calendar life cycle. In power systems applications, there is a possibility of having pulse load, causing severe thermal and power disturbances in the power system and also in micro-grid applications. The UC, because of its fast response to power balancing and leveling with an appropriate control system in place, can help in overcoming these challenges. A major drawback of the UC is its high self-discharge rate and cost.

The operation of the super magnetic energy storage (SMES) is based on the principle of electrodynamics. By circulating current in the superconducting

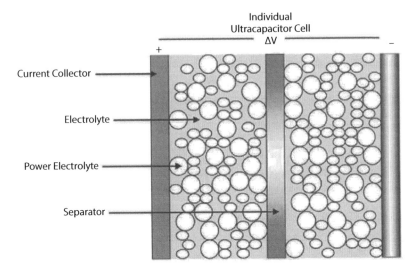

Figure 1.6 Doubled layer supercapacitor [13].

coil while in charging mode, energy is stored in the magnetic field of the SMES device, therefore reducing the superconductivity at low temperatures. The superconducting material present in the SMES is cryogenically cooled. The implication of this is that stored energy can be released back to the network by discharging the coil (discharging mode). A major concern of this storage technology is heat generated by ohmic losses thereby causing thermal instability. The high cost of installation is also a drawback of this technology.

C. Thermal storage systems

The thermal energy storage (TES) systems store energy as either heat or ice, which can be released later when needed. It is an alternative technology to replacing the usage of fossil fuels and meeting the demand for sustainable energy regulations. TES can also be used for applications such as heating or cooling systems in industrial and residential sectors, power generation, and load shifting.

TES consists of a thermal storage tank, a heat transfer medium, and a containment control system. Using distinctive technology, heat is stored and kept in an insulated reservoir. Figure 1.7 depicts a diagrammatic representation of a thermal energy storage system. Stored heat energy is initially transferred and then transformed into electricity using a heat engine cycle [15]. The containment control system is required for the general operation of the reservoir and to monitor the heat transfer medium. It is important to note that TES systems have a very low cycle efficiency in the range of 30-30%; however, this storage technology has a high daily self-discharge

Current Development in Energy Storage Technologies 11

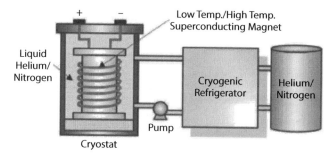

Figure 1.7 Schematic of a SMES [14].

and energy rate, environmental friendliness, and the low-cost initial capital, which are some of the advantages of the TES. The schematic of a TES is shown in Figure 1.8 below.

TES systems are broadly classified into two groups based on operating temperature. These are low-temperature TES and high-temperature TES. Low-temperature TES systems are made of cryogenic energy storage and auriferous low-temperature storage typically operating at temperatures below 200 degrees Celsius and find their applications in water heating and solar cooking [17]. The high-temperature TES can be further subdivided into three categories, namely: latent heat system (LHS), absorption and adsorption system (AAS), and sensible heat (SHS). The schematic of a TES is shown in Figure 1.8.

Latent heat storage systems (LHS) cannot be sensed by a change in temperature. An important feature of the LHS is thermal conductivity, improved with metal fillers, metal matrix structures, aluminium shavings, and paraffin. The enthalpy and density of the system is a determinant

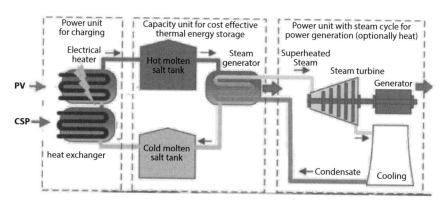

Figure 1.8 Schematic of a TES [16].

factor of its storage capacity while the medium of storage can be organic, inorganic, or bio-based.

The absorption and adsorption system (AAS) is based on a physicochemical process. This is also referred to as a thermo-chemical storage system. With this storage technology, heat is not stored directly. High energy density is the main advantage of the AAS.

The sensible heat (SHS) can regulate the mass medium and specific heat capacity used for determining the capacity of the storage system during the storage operation processes. The medium of storage used for SHS can be categorized into different forms such as liquid (thermal oil, molten salt, and water), liquid with solid filter material (stone/molten salt), and solid (concrete, metal, and ground). A main drawback of the SHS is the size requirement.

D. Chemical energy storage (CES)

This energy storage system is suitable for storing significant amounts of energy for long periods. In this system, the energy is stored in the chemical bonds between molecules and atoms that are released through the transfer of electrons to directly produce electricity. The most popular sources of chemical fuels for generating electricity and energy transportation systems are coal, ethanol, propane, hydrogen, gasoline, diesel, and liquefied petroleum gas (LPG). The focus of the CES system is hydrogen technology because of its remarkable ability to act as fuel and also store large amounts of electrical energy.

Hydrogen fuel cell (HFC) is a promising alternative for emission-free electricity generation, and its application can be extended to distributed generation and the automobile industry. In comparison to other hydrocarbon fuels, it releases only water vapor into the environment when burned, though it burns faster and contains considerable chemical energy per mass. Its high energy density by weight and low energy density by volume as well as the environmental friendliness of hydrogen has made it become a focus for an energy storage solution. HFC as a medium of energy storage is also very cost-effective considering that hydrogen is readily available.

The application of hydrogen to energy storage systems can be divided into four categories, namely: hydrogen liquefaction, hydrogen adsorption on carbon nanofibers, metal hydrides absorption, and hydrogen pressurization. The pressurization of hydrogen is dependent on the high permeability of materials. A pressure of about 200-250 bars can be stored in a steel tank. However, this is only possible when the ratio of stored hydrogen per unit weight is low since the storage efficiency depends on the increase in temperature. Metal hydride energy storage systems are dependent on the properties of hydrogen absorption of the chemical compounds present. The advantages of this storage system include compatibility and low pressure since most hydrides have higher rates of absorption equivalent to the volume of hydrogen stored to the volume of metal used.

Hydrogen storage technology is preferred for load-shifting applications; however, this technique is costly and its efficiency is the most critical criterion to develop this technology [18].

E. Electrochemical energy storage

In electrochemical storage systems, chemical energy is converted into electrical energy. The process of energy conversion is completed by chemical reactions, while the energy is stored as an electric current for a specific voltage and time. Electrochemical energy storage solutions are the largest group of energy storage devices available. Conventional flow and rechargeable batteries are two technologies for storing energy in electrochemical form. Though there is minimal maintenance needed for batteries, the life expectancy and energy are reduced as a result of a chemical reaction.

Electrochemical storage devices have the advantage of being readily available in different sizes. Examples of the different electrochemical storage devices include lithium-ion, sodium-sulphur, nickel-cadmium, Lead-acid, and flow batteries.

F. Hybrid energy storage

As the name implies, a hybrid energy storage system integrates two or more energy storage technologies, capitalizing on the advantages each of them provides to obtain an excellent solution for a particular application. Since it is practically impossible for single storage technology to have all the advantages, hybrid energy storage is becoming more appealing for modern-day applications such as in micro-grids.

Fast response at high rates for a short duration can be obtained by implementing high-power ESS, while slow response applications for an extended time are achieved with high-energy devices. Micro-grids and power systems need an energy storage solution that can combine the characteristics of a high power and high energy storage system to reduce the problems associated with power quality and improve the stability of the power system. Hybrid energy storage solutions have more complicated control structures than those of single energy storage systems. Some of the possible configurations for hybrid energy storage solutions include the following:

- CAES + SMES
- CAES + FES

- CAES + UC
- CAES + Battery
- Pumped hydro + battery
- Pumped hydro + UC
- Pumped hydro + SMES
- Battery + SMES
- Battery + UC

1.3 Energy Storage and Smart Grids

With the ever-increasing growth in the global population comes a corresponding incremental demand for electrical energy. This has become a challenge not only for the generation of electrical energy but also for its distribution. The growing demand for electrical energy has increased the complexities associated with power grids by increasing the requirements for greater reliability, security, efficiency, and environmental and energy sustainability concerns.

Renewable energy sources (RES) have been proposed as suitable alternatives to conventional fossil power plants in meeting the rising demand for electrical energy. The integration of RES into the power grid has been a source of concern for system operators. This is because of the intermittent nature of such RES and their impact on grid flexibility and balance.

Grid flexibility is the ability of the power system to quickly respond to conditions of unpredicted variations. With conventional power plants, this requirement is made possible by managing the power plant output by rapidly ramping up and down to match the variable and non-predictable electric load. However, with grid integration of RES, the unpredictability of the power system has shifted from the demand side to the generation side and in response to adjustment, flexibility agents have conversely migrated to the demand side. Nevertheless, the current power systems can still make room for flexibility by managing the output of the power plants; with RES this translates into curtailment, therefore, losing clean energy.

To improve power systems and grid flexibility, especially with the grid integration of RES, the concept of smart grids has been introduced. A smart grid is simply a grid that uses modern communication technologies to integrate various elements of an energy system, such as generation and demand, thereby ensuring a balance on both sides. Smart grids ensure that RES can be adequately and optimally fed into the grid. Such a grid can digitally gather, distribute and use provided information on all participants such as suppliers and consumers to improve the reliability, efficiency, and economics of the electricity services [19].

The development of a smart grid comes with the difficult challenge of ensuring a balance between the various variables in connection with dynamic load control powered by the ever-increasing penetration of RES. Installing energy storage solutions can help in achieving the balance between these variables. Energy storage is a very important component of a smart grid and the different ESS available have been well documented in Section 1.2 of this chapter. The choice of ESS for a smart grid is dependent on the number of charging/discharging cycles and the duration of such operations.

ESS plays a complementary role in meeting the goals of an efficient smart grid. The benefits of a smart grid to the power system include less costly interruptions, improved reliability of the power system, an increase in efficiency of power delivery with lower losses in distribution as well as deferred capital expenditure on transmission and generation assets. Energy storage helps the smart grid actualize these merits by negating the need for extra peaking generation by deferring load transmission and distribution upgrades, reducing the transmission congestion fees in deregulated markets by the addition of energy storage to distribution substations, and provision of load capabilities that can help in improving the intermittency of renewable energy sources.

ESS incorporated into the smart grid can also bring cost-saving benefits to the residential consumer from cost savings from peak load management and energy efficiency. On the other hand, the smart grid also provides an opportunity for dispatch of storage units and load control, making RES more valuable to the grid.

A major advantage of a smart grid is the allowance for a wider range of technologies. The use of a fast response ESS such as a flywheel in a smart grid can potentially improve the overall power quality and reliability of the grid by quickly responding to intermittency issues.

Though ESS can facilitate improving power quality and the overall reliability of the grid, it is practically impossible for the ESS to deliver these advantages at every given time. Therefore, the smart grid makes a "smart decision" on the most technical benefit to the grid at any given time, co-optimizing between revenue streams such as ancillary markets and arbitrage [20].

1.4 Energy Storage and Micro-Grids

Micro-grids (MG) are symbols of controllable electric entities containing different loads of distributed energy sources. All typical micro-grids produce electricity from two or more sources, and at least one of such sources is a renewable energy source. Micro-grids are also susceptible to stability and power balance, hence the integration of ESS with a micro-grid is of great benefit to the micro-grid, operating either in an islanded mode or grid-connected mode.

An ESS can be added at various positions in the MG depending on the purpose. The ESS can undertake peak shifting and load leveling functions while acting as a load, supporting peak demand. With the integration of RES, fluctuation mitigation can be performed by the ESS. For an MG with solar photovoltaic and wind power generation, the ESS can enhance the low voltage ride-through capability. It is important to note that the configuration of an ESS in a MG is dependent on the application. For MGs with renewable energy penetration, the ESS can be configured as a distributed ESS or an aggregated ESS.

1. Distributed ESS

In the distributed ESS configuration, the ESS units are coupled directly to the individual distributed generators with different interfaces. Distributed ESS for micro-grids can be further configured to be positioned either on the generator side or the load side of the MG. The ESS positioned on the generator side helps in the generation of smooth output power while on the load side, it helps in the reduction of load variation and energy management. For the generator side distributed ESS, it is usually connected to the DC link of the renewable energy generation unit. A major advantage of the distributed ESS is easy maintenance, optimized cost, and efficiency.

For the load-side distributed ESS, these are usually connected to the local load as the name implies. Variations in load configuration can have a significant impact on the operation of a MG. Therefore, the distributed generators need to be able to follow load change when there is a rapid change in load. If the loads cannot be matched, then the MG becomes unstable. Load following is not very energy efficient since the distributed generators cannot always operate at rated conditions. Therefore, the deployment of distributed ESS can help in smoothing load variation.

2. Aggregated ESS

The aggregated ESS configuration is popular with a lot of MG projects. This is usually a big energy storage facility that offers dedicated housing in the micro-grid. In the aggregated ESS configuration, all the ESS are connected to the terminal of the MG as one aggregated ESS. With the aggregated ESS, there is a large capacity to store a vast amount of energy. Such a storage system may consist of several storage units. Examples of an aggregated

ESS is a battery energy storage system (BESS) comprising several battery packages, or a FES made up of several flywheel units. In comparison to the distributed ESS, the aggregated ESS is a better option for suppressing power fluctuations in the MG [21] since the aggregated ESS can stabilize the entire MG, unlike the distributed ESS which works specifically with its own distributed generator.

1.5 Energy Storage Policy Recommendations

Investment tax credits (ITCs) are a very effective way to reduce capital costs and limit exposure to technological and capital risk. This can be applied to energy storage solutions. The implementation of short-term tax policies can help in addressing the high capital cost associated with different energy storage options. ITCs can promote a rapid increase in the storage capacity for services such as frequency regulation. It can also promote the widespread use of technology by boosting market demand and promoting the cost-competitiveness of new technologies against conventional ones. Also, the renewable energy markets have clearly shown the importance of giving tax break incentives, including ITC regimes [22].

ITC, if effectively implemented, can promote the expansion of energy storage in the short run while simultaneously accelerating the reduction of capital costs in the long run. To reach the full potential of an effective smart grid, the importance of an energy storage market characterized by low costs and stable supply cannot be over-emphasized. However, it is important that any ITC program introduced should be able to prevent market volatility in the long term.

Technology risks can be an obstacle to energy storage technologies, and different technically viable energy storage technologies have experienced a delay in deployment because of reluctant developers who are not willing to be the first to carry out such project systems. Pilot and demonstration projects are therefore very important to showcase the practicality of new storage technologies, and successful demonstrations of such technology can help in obtaining private investor funding for the large-scale development of such energy storage systems. It is therefore important for governments across the globe to invest in research and development projects to encourage the development of low-cost effective energy storage solutions. Research and development can reduce the relatively high capital costs of ESS by enhancing the possibility of revenue streams from ESS through increased applications and showing how viable the next generation of energy storage technologies can be.

Also, market formation and support for energy storage solutions are important for the growth of this sector. The various government regulatory agencies in charge of energy should be encouraged to support the deregulation of the energy and power markets, taking necessary steps to ensure stability and market development. Various rules and policies can be developed that monitor storage operations, and evaluate and provide ancillary storage services by providing cost of service rates to owners of storage assets to secure revenue streams and access to capital.

Regional transmission organizations and utilities should be encouraged to participate in the wholesale market, providing access to storage owners by the creation of interconnection standards and compensation schemes for the owners of small-scale storage units. This will in turn boost market competition. By deregulating utilities, and energy markets and supporting greater regulatory market structures, storage owners will have access to the energy markets. This will help in the diversification of the energy markets concerning the services available for exchange and eligible participants.

1.6 Energy Storage: Challenges and Opportunities

Energy storage systems are vital in the ever-increasing demand for energy and large-scale penetration of renewable energy sources in the march towards utilization of low carbon energy sources. However, the large-scale deployment of energy storage technologies still faces technical and economic as well as regulatory challenges.

Currently, policy and regulatory uncertainties, high technological costs, and the energy market structures are some of the challenges facing the large-scale deployment of ESS [23]. The pricing of energy storage is dependent on the energy market structures. By making the storage market competitive, profits can be maximized by storing energy at low prices and releasing it when prices are higher. A competitive energy market inclusive of storage will eliminate investment distortions, thereby encouraging investments in storage capacity. It is also vital to propose clarifications for energy storage systems with active participation from all classes of users to support a dynamic market structure. Governments should aim to facilitate and provide incentives in the energy storage sector considering the regulatory and economic context.

The development of ESS requires innovative and cutting-edge breakthrough research focused on developing more affordable technologies, low cost, longer life span, capacity, and high security for energy storage solutions [24]. Such research should be focused on the simulation of energy storage and operation optimization for multiple applications that can support the application of such storage technologies from a theoretical point of view. The development of demonstration projects that can promote the commercialization and industrialization of such energy storage technologies should be encouraged. There is also the need to encourage collaboration between various research groups focusing on ESS and industry counterparts.

Batteries represent a significant portion of ESS; therefore there is a need to develop strict recycling regulations for batteries aimed at materials recovery. There are lots of inherent opportunities in the local recycling and repurposing of batteries. However, there is a need for research on this to improve the economics of it [5].

1.7 Practical Implementations of Energy Storage Technologies

Energy storage remains an important component of the electricity network, especially for micro-grids and renewable energy sources. Storage systems play the same role across all applications: absorbing generated energy and discharging it later. Energy storage applications include:

- Spinning reserve
- Frequency regulation
- Peak shaving
- Voltage support
- Power quality and
- Capacity firming.

It is important to stress that no two energy storage applications are the same. Different companies offer different technologies for energy storage in live grids across the globe. It is estimated that the global installed storage capacity will expand by at least 56% over the next five years, reaching over 270 GW at the end of 2026 [5]. The bulk of energy storage technologies will be in the form of utility-scale batteries. According to the International Energy Agency (IEA), pumped storage hydro is expected to provide about 42% of global electricity storage capacity with over 40 GW of expansion in the next five years [5].

There are a few examples of practical implementation of energy storage solutions globally. The Golden Valley Electric Association (GVEA) in

Alaska uses a nickel-cadmium battery system to supply 27 MW of power for 15 minutes or 46 MW for 5 minutes, allowing ample time for local generation to come online while the Swedish utility company, Falbygdens, utilizes a battery energy storage system to locally produce energy from wind turbine by creating a storage capacity of 75 kilowatts in cycles that last up to 60 minutes. This helps in providing stability to the grid and balancing peak loads during the day. The energy storage market in Germany is growing at a remarkably fast pace. The German utility and automation company, Steag, has inaugurated a 90 MW energy storage system. This project consists of six 15 MW Nidec ASI storage systems, with an investment of USD 100 million. In February 2018, Enel signed an agreement with the German wind energy company, Enertrag AG, and Swiss energy storage solutions company, Leclanché SA, to build and manage a 22 MW lithium-ion battery storage plant in Cremzow in the German state of Brandenburg. With Germany's ambitious aim of 65% renewable energy deployment by 2030, it is expected that the demand for energy storage will increase at a considerable rate during the forecast period.

1.8 Conclusions

This chapter has discussed the operating principle and characteristics of different energy storage technologies suitable for different power systems applications. Energy storage remains a very important component of the power system with the large-scale penetration of renewable energy into the grid to meet the rising energy demand. The choice of energy storage solution is application specific. ESS also plays a very important role in smart grids. Though ESS can facilitate improving power quality and the overall reliability of the grid, it is practically impossible for the ESS to deliver these advantages at every given time. Therefore, the smart grid makes a "smart decision" on the most technical benefit to the grid at any given time, co-optimizing between revenue streams such as ancillary markets and arbitrage. Though the numerous advantages of ESS have been listed, there are still technical and economic challenges mitigating against the large-scale deployment of ESS.

References

1. Salawitch, Ross J., Timothy P. Canty, Austin P. Hope, Walter R. Tribett, and Brian F. Bennett. *Paris climate agreement: Beacon of hope*. Springer Nature, 2017.

2. DeConto, R.M., Pollard, D., Alley, R.B., Velicogna, I., Gasson, E., Gomez, N., Sadai, S., Condron, A., Gilford, D.M., Ashe, E.L. and Kopp, R.E., 2021. The Paris Climate Agreement and future sea-level rise from Antarctica. *Nature*, 593(7857), pp.83-89.
3. IEA (2021), Renewables 2021, IEA, Paris, https://www.iea.org/reports/renewables-2021.htm, 2021.
4. Handschin, Edmund, ed. *Power system applications of the modern battery storage*. Univ., 2004.
5. IEA (2021), "How rapidly will the global electricity storage market grow by 2026?" IEA, Paris, https://www.iea.org/articles/how-rapidly-will-the-global-electricity-storage-market-grow-by-2026
6. Seferlis, Panos, Petar Sabev Varbanov, Athanasios I. Papadopoulos, Hon Huin Chin, and Jiří Jaromír Klemeš. "Sustainable design, integration, and operation for energy high-performance process systems." *Energy* 224 (2021): 120158.
7. Choi, Wooyoung, et al. "Reviews on grid-connected inverter, utility-scaled battery energy storage system, and vehicle-to-grid application-challenges and opportunities." *2017 IEEE Transportation Electrification Conference and Expo (ITEC)*. IEEE, 2017.
8. Luo, Xing, et al. "Overview of current development in electrical energy storage technologies and the application potential in power system operation." *Applied Energy* 137 (2015): 511-536.
9. Cheng, Jie, and F. Fred Choobineh. "A comparative study of the storage assisted wind power conversion systems." *2017 6th International Conference on Clean Electrical Power (ICCEP)*. IEEE, 2017.
10. Molina, Marcelo G. "Energy storage and power electronics technologies: A strong combination to empower the transformation to the smart grid." *Proceedings of the IEEE* 105.11 (2017): 2191-2219.
11. Berrada, Asmae, Khalid Loudiyi, and Izeddine Zorkani. "System design and economic performance of gravity energy storage." *Journal of Cleaner Production* 156 (2017): 317-326.
12. Hadjipaschalis, Ioannis, Andreas Poullikkas, and Venizelos Efthimiou. "Overview of current and future energy storage technologies for electric power applications." *Renewable and Sustainable Energy Reviews* 13.6-7 (2009): 1513-1522.
13. Wang, Guoping, Lei Zhang, and Jiujun Zhang. "A review of electrode materials for electrochemical supercapacitors." *Chemical Society Reviews* 41.2 (2012): 797-828.
14. Molina, Marcelo G. "Distributed energy storage systems for applications in future smart grids." *2012 Sixth IEEE/PES Transmission and Distribution: Latin America Conference and Exposition (T&D-LA)*. IEEE, 2012.
15. Chen, Haisheng, et al. "Progress in electrical energy storage system: A critical review." *Progress in Natural Science* 19.3 (2009): 291-312.

16. Kramer, Susan. "Morocco Pioneers PV with Thermal Storage at 800 MW Midelt CSP Project." *SolarPACES, April* 25 (2020).
17. Gil, Antoni, *et al.* "State of the art on high-temperature thermal energy storage for power generation. Part 1—Concepts, materials, and modelization." *Renewable and Sustainable Energy Reviews* 14.1 (2010): 31-55.
18. Kousksou, Tarik, *et al.* "Energy storage: Applications and challenges." *Solar Energy Materials and Solar Cells* 120 (2014): 59-80.
19. U.S. Department of Energy, 2012. Smart Grid/Department of Energy.
20. Roberts, Bradford P., and Chet Sandberg. "The role of energy storage in development of smart grids." *Proceedings of the IEEE* 99.6 (2011): 1139-1144.
21. Wei L, Joos G. Performance comparison of aggregated and distributed energy storage systems in a wind farm for wind power fluctuation suppression. In: Power engineering society general meeting. IEEE. Tampa, FL, USA: 2007.
22. Sawin, Janet, and Christoper Flavin. "Policy lessons for the advancement & diffusion of renewable energy technologies around the world." *trabajo presentado en la International Conference for Renewable Energies, Bonn.* 2004.
23. Bhatnagar, Dhruv, *et al.* Market and policy barriers to energy storage deployment. No. SAND2013-7606. Sandia National Lab. (SNL-NM), Albuquerque, NM (United States); United States., Washington, DC, 2013.
24. S. Dhundhara, Y. P. Verma, and A. Williams, "Techno-economic analysis of the lithium-ion and lead-acid battery in microgrid systems," *Energy Convers. Manag.*, vol. 177, pp. 122–142, 2018.

2

A Comprehensive Review of the Li-Ion Batteries Fast-Charging Protocols

Talal Mouais[1] and Saeed Mian Qaisar[1,2]*

[1]*Electrical and Computer Engineering Department, Effat University, Jeddah, Saudi Arabia*
[2]*LINEACT CESI, Lyon, France*

Abstract

One of the significant drawbacks of renewable energy sources, such as solar and wind, is their intermittent pattern of functioning. One promising method to overcome this limitation is to use a battery pack to enable renewable energy generation to be stored until required. Batteries are known for their high commercial potential, fast response time, modularity, and flexible installation. Therefore, they are a very attractive option for renewable energy storage, peak shaving during intensive grid loads, and a backup system for controlling the voltage drops in the energy grid. The lithium-ion (Li-Ion) is considered one of the most promising battery technologies. It has a high energy density, fair performance-to-cost ratio, and long life compared to its counterparts. With an evolved deployment of Li-Ion batteries, the latest trend is to investigate the opportunities of fast Li-Ion battery charging protocols. The aim is to attain around the 70-80% State of Charge (SoC) within a few minutes. However, fast charging is a challenging approach. The cathode particle monitoring and electrolyte transportation limitations are the major bottlenecks in this regard. Additionally, sophisticated process control mechanisms are necessary to avoid overcharging, which can cause a rapid diminishing in the battery capacity and life. This chapter mainly focuses on an important aspect of realizing the effective and fast-charging protocols of Li-Ion batteries. It presents a comprehensive survey on the advancement of fast-charging battery materials and protocols. Additionally, the state-of-the-art approaches of optimizing the configurations of concurrent fast-charging protocols to maximize the Li-Ion batteries life cycle are also presented.

---

*Corresponding author: smianqaisar@cesi.fr

Sandeep Dhundhara, Yajvender Pal Verma, and Ashwani Kumar (eds.) *Energy Storage Technologies in Grid Modernization*, (23–70) © 2023 Scrivener Publishing LLC

Keywords: Lithium-ion battery, renewable energy, lifetime of a battery, control system, fast charging, protocols, battery materials, system optimization

2.1 Introduction

A battery is a very significant energy storage device. Each battery converts chemical energy to electricity, and it contains three main parts: the electrolyte, electrodes, and a separator. A lithium-ion battery is considered one of the most promising battery technologies because it has a high energy density, a low falling cost, and a long lifetime [1]. A comparison among the energy densities of various rechargeable batteries is displayed in Figure 2.1. It shows the outperformance of Li-Ion batteries over their counterparts.

The Li-Ion battery has many important applications. For instance, in the medical field, Li-Ion batteries are used in implantable cardiac pacemakers as shown in Figure 2.2. A device sends an electrical pulse to the heart at the right strength to stimulate it at the right pace.

Also important is the use of Li-Ion batteries for energy storage in the space exploration as shown in Figure 2.3. In 2010, NASA replaced all Ni-H2 batteries in the International Space Station (ISS) with Lithium-ion (Li-Ion). Li-Ion battery is also used to power the spacesuits of astronauts.

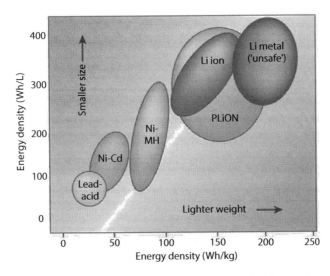

Figure 2.1 The energy densities and specific energies of several rechargeable batteries.

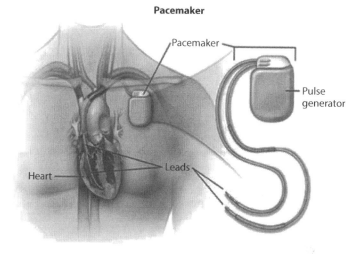

Figure 2.2 Cardiac pacemakers. Source: Adapted from [2].

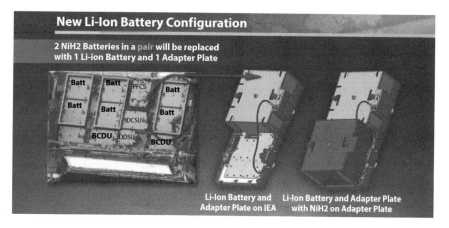

Figure 2.3 Li-Ion batteries used in the international space station. Source: Adapted from [3].

The Li-Ion battery shows promise as a technology that can store energy for renewable energy technologies, such as wind and solar. A major challenge of renewable energy is that it supplies energy intermittently; the electrical output varies depending on the sun's intensity, wind speed, and other factors. Overcoming the intermittent energy supply problem, Li-Ion batteries can store the energy and supply it to the grid during energy supply shortages. The basic battery working principle during discharge/charge cycle is shown in Figure 2.4.

The development of electric vehicles has resulted in a greater need for batteries with large storage capacities and quick-charging capabilities. For these vehicles, lithium-ion batteries have shown to be one of the most promising energy storage systems. In comparison to other battery types, Li-Ion batteries offer a high power and energy density. The battery offers a number of advantages over other types of batteries. Li-Ion batteries, for example, may be charged from a variety of sources, including the AC grid (through a power electronic converter). Electric motors that are powered by an engine or that are in regenerative mode can also be used to charge Li-Ion batteries.

The term "battery" refers to a device that produces electrical energy from chemical reactions. Different types of batteries exist using different chemicals. Usually, a battery cell has two chemicals with different loads connected with a negative electrode and a positive electrode. The cathode supplies a current of electrons to a connected appliance while the anode accepts the electrons. Figure 2.4, adapted from [4], shows the basic battery working principle during discharge/charge.

Energy storage for renewable energy sources requires rechargeable batteries [5]. The available rechargeable battery technology includes lead-acid, Nickel-Cadmium, Li-Ion batteries, and Sodium-ion batteries, each of which has advantages and disadvantages. When selecting a rechargeable battery for a renewable energy system, the significant factors to consider include energy density, capacity, cell voltage, efficiency, and lifetime.

The amount of energy held in a battery per unit volume is referred to as energy density. It can be stated either gravimetrically or volumetrically. The quantity of energy contained in a battery in relation to its weight, given

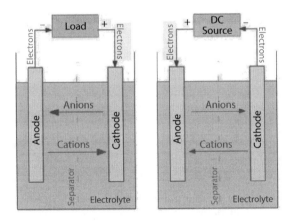

Figure 2.4 Basic battery working principle during discharge/charge.

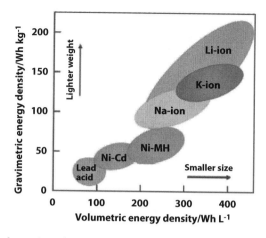

Figure 2.5 The volumetric and gravimetric energy densities for various batteries.

in Watt-hours per kilogram (W-hr/kg), is referred to as gravimetric energy density. Volumetric energy density, on the other hand, refers to energy measured in Watt-hours per liter (W-hr/l) [6]. The Ragone figure, shown in Figure 2.5 [7], demonstrates the volumetric energy density vs. the gravimetric energy density of several battery systems.

This study will focus on one of the most critical aspects of BMSs: the implementation of successful Li-Ion battery fast-charging methods. A complete overview of the evolution of rapid charging battery materials and processes will also be presented. We will also go through the most up-to-date methods for maximizing the life cycle of Li-Ion batteries by optimizing the configurations of concurrent fast-charging protocols.

2.2 The Literature Review

Climate change has become a menace in recent years, making governments and environmental agencies suggest strategies to solve the problem. The transportation sector is considered one of the key contributors to pollution leading to climate change. Pollution comes from combustion engine vehicles that primarily rely on carbon-based fuels that emit high levels of greenhouse gases to the atmosphere. The development of lithium-ion-powered electric vehicles is among the solutions to combating global climate change. Car manufacturers have made efforts to include EVs in their range of vehicles. However, recent studies have identified significant gaps in the level of acceptance of these vehicles, especially those that are

28 Energy Storage Technologies in Grid Modernization

battery-powered [8]. Vehicle users have always pointed out long charging times and range anxiety as among the core reasons they are not ready to leave their combustion engine vehicles, which they can refill easily within a short time [9]. This has made fast charging capabilities the primary focus for battery developers and electric vehicle manufacturers. Tomaszeweska *et al.* [8] recognize that manufacturers have made efforts to create fast-charging batteries. However, charging at fast rates to reduce the overall charging time causes battery deterioration due to increased degradation.

Additionally, fast charging leads to increased heat generation, making it difficult to remove. These factors have led to a lot of safety and degradation concerns that most studies are yet to address. It is evident that existing literature has significant knowledge gaps, despite addressing the need for fast-charging protocols for batteries used in electric vehicles.

2.2.1 Overview of Lithium-Ion Battery Working Principle

The outer shell of the lithium atom has just one electron. As a result, it has the greatest chance of losing an electron (pure lithium is a very reactive metal, even it reacts with water and air). Lithium, on the other hand, is stable in its metal oxide state. The occurrence of Lithium in the periodic table is shown in Figure 2.6. The following are the basic phenomena of power production in Li-Ion batteries:

- The lithium atom is separated from the lithium metal oxide.
- Directing an electron that has been lost from a lithium atom to an external circuit.

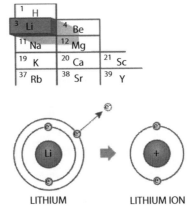

Figure 2.6 Lithium in the periodic table.

* During Charging Lithium-Ion battery:

When a Li-Ion battery is connected to a power source, the positive side attracts and removes electrons from the Lithium oxide. These electrons pass through the external circuit because they cannot reach the graphite layer through the electrolyte. Meanwhile, the positively charged Li-Ion is drawn to the negative terminal and passes through the electrolyte. When Li-Ion reaches the graphite layer, it is trapped and stored between the layers of carbon graphite. Lithium will be able to act as a wedge between each layer. Intercalation is the term for this occurrence. A Li-Ion battery is seen charging in Figure 2.7 from [10].

* During discharging Li-Ion battery:

During discharge, the Li-Ion wants to return to its stable state, so the lithium ions go through the electrolyte to the metal oxide. On the other hand, the electron will go to the external circuit. Figure 2.8 from [10] shows the Li-Ion battery during discharging.

The liquid electrolyte will dry up if the cell's internal temperature rises owing to abnormal circumstances. This might result in a fire or an explosion. An insulator layer called a separator is put between the electrodes to prevent this [11]. Because of its microporosity, the separator is permeable to lithium ions. The electron, on the other hand, will not be able to travel through it. The electrochemical process in a Li-Ion battery is illustrated with the help of Figure 2.9.

SEI refers to a significant occurrence that occurs in a Li-Ion battery when it is charged for the first time (solid electrolyte interface). Solvent molecules in the electrolyte cover the lithium ions as they flow through the electrolyte; when they reach the graphite, the lithium ions in the solvent molecules react with the graphite and produce an SEI layer. This layer is

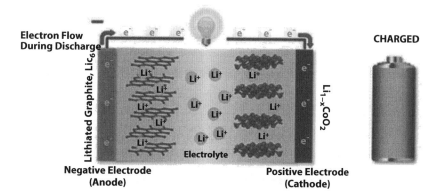

Figure 2.7 Li-Ion battery during charging.

30 Energy Storage Technologies in Grid Modernization

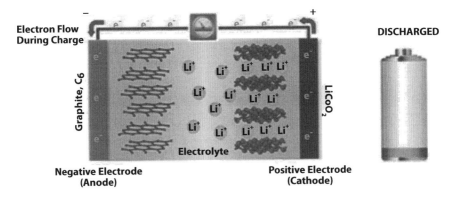

Figure 2.8 Li-Ion battery during discharging.

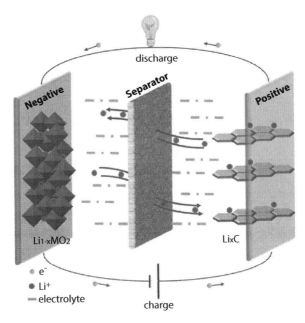

Figure 2.9 Illustration of the electrochemical process in a Li-Ion battery [12].

important because it avoids direct contact between the electron and the electrolyte, which protects the electrolyte from deterioration. The process of creating the SEI uses 5% of the life of a Li-Ion battery [13]. The remaining 95% of Li ensures that the battery continues to function.

The dimensions and chemistry of the SEI layer can be optimized to improve the Li-Ion battery's performance. We do not discuss this further

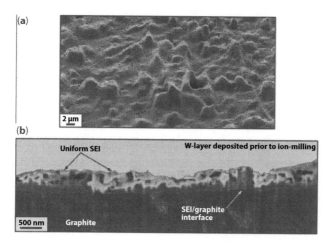

Figure 2.10 Scanning electron microscope image shows the SEI layer's morphology formed on graphite at 60 °C after the first cycle.

here. Figure 2.10, adapted from [14], shows a scanning electron microscope (SEM) image of the SEI layer (Figure 2.10a) and its cross-sectional microstructure (Figure 2.10b).

2.2.2 Principles of Battery Fast-Charging

The deployment of electric vehicles depends on the availability of batteries with efficient life and charging capacities. An ideal battery is expected to exhibit a long lifetime, with sufficient life and power densities. Such batteries would allow EV users to achieve long-range travel and fast recharge despite location and weather. However, it is not easy to achieve all these characteristics in a battery, leading to trade-offs during the manufacturing process [15]. For instance, thicker electrodes have been used to achieve high energy density. However, the electrodes are prone to extreme gradients and concentration due to fast charging [16]. Battery operation depends on effectively combining the physical properties of manufacturing materials with devices that have temperature-dependent behaviors. The operating temperature has been identified as the major barrier to fast charging as rapid temperature increases pose severe safety risks. Therefore, charge rates are always kept low to maintain the required ambient temperatures [15]. The lithium-ion batteries, the lithium plate's risk is significantly high when temperatures decrease. This affects the overall capacity retention of the battery. The minimum temperature threshold at which the lithium plate is exposed to risk depends on various factors. Common factors that determine the level of risk that the plate is exposed to risk include cell age, parameters, and the C-rate. Various scholars have cited 25 degrees Celsius as the minimum temperature for this [17].

Fast-charging batteries are preferred for two primary reasons [18]. First, the batteries reduce the overall charging time. Second, fast-charging batteries reduce degradation levels and ensure that heat generation remains as low as possible since these are the two main challenges that are preventing electric vehicles from becoming fully adopted in the modern transport market [49].

The constant current-constant voltage (CCCV) charging procedure is used by lithium-ion battery chargers [18]. There are two steps to this charging technique. The battery is first charged at a steady current until it meets the manufacturer's limit. The constant voltage stage follows, with the voltage remaining at the upper limit to guarantee that charging does not continue until the charging current falls below the pre-determined battery cut-off value. The CCCV protocol can be used to accomplish quick charging. However, because the constant current must be set at a high level in the initial part of the technique, it has undesirable consequences, such as strong polarization. Furthermore, strong polarization can increase the charging voltage to its top limit, causing the charging mode to flip from the first to the second stage. As a result, the CCCV protocol is used to charge the Li-Ion battery at a low rate, requiring additional charging time, because attaining quick charging with this protocol results in strong polarization, which can lead to lithium plating.

Fast-charging protocols have been suggested as the solution to high-voltage current created when the charging current is high. The proposed protocols use decayed charging current profiles to ensure that the charging voltage does not increase during fast-charging processes. The multi-stage constant current protocol is one of the recommended procedures (MSCC). The protocol has the potential to minimize total charging time and heat, hence enhancing overall charging energy efficiency [19]. Up to four layers of constant current charging are used in the multi-stage constant current technique. Because a large current may be used during the first charging stage and a low current can be used during the final charging stage, it is simple to avoid polarization during charging when these different levels are employed. The variable current decay (VCD), linear current decay (LCD), and constant voltage-constant current-constant voltage (CV-CC-CV) methods have also been proposed as fast-charging protocols [20]. These procedures follow the same guidelines as the MSCC.

2.2.3 Multi-Scale Design for Fast Charging

As noted in several studies, the degree and mode of degradation resulting from fast charging depends on battery material components [8]. The components include electrode and electrolyte properties, operational conditions, manufacturing processes, and pack design [8]. Despite these challenges, the multi-scale design provides significant opportunities for developing efficient and fast-charging batteries.

2.2.4 Electrode Materials

Electrode materials are crucial to the cell design of any form of battery. According to Tomaszeweska *et al.* [8], selecting suitable electrolyte and electrode combinations in the batteries' physics to achieve high capacities is one of the most challenging aspects of the manufacturing process. Extensive research has been conducted in developing lithium fast-charging anode materials that can limit the possibility of lithium dendrite formation as temperatures change during charging and after [8]. There is a significant success in these researches as some materials, such as carbon-based alternatives like graphite and alloy composites, have proved to be effective.

Traditional graphite anodes are chosen for generating maximum cell energy density because they have a high potential for lithium/lithium-ion production. The anodes, on the other hand, are prone to lithium plating. As a result, changing the anode material is the most effective way to improve the quick-charging capabilities of Li-Ion batteries. The reversible capacity of surface-engineered graphite with a 1 wt percent Al_2O_3 covering demonstrates this. At a rate of 4000 mAg1, graphite reaches a reversible capacity of 337 mAh g-1 [8]. Because they do not have lithium plating issues, LTO lithium-titanate-oxide (LTO) materials are also attractive for batteries with quick-charging capabilities and extended lives. However, the materials' working potentials are exceedingly high, resulting in poor cell voltage and energy density. LTO has undergone some changes, including the addition of carbon sources and a coating. The main goals of these changes are to improve electric conductivity and active powder content. "Vanadium disulfide flakes with Nano layered titanium disulfide coating as cathode materials in lithium-ion batteries," according to "Vanadium disulfide flakes with Nano layered titanium disulfide coating as cathode materials in lithium-ion batteries."

Because vanadium disulfide (VS2) is very conductive, replacing cobalt oxide in the cathode will result in quicker charging. Coating vanadium disulfide (VS2) with TiS2 also serves as a buffer layer, holding the VS2 material together and giving mechanical support.

2.2.5 Fast-Charging Strategies

Many material solutions suggested in the study show promising results. However, most of these solutions are not expected to reach the broader market level in the future, considering that more countries are adopting electric vehicle technologies. Therefore, more researchers have shifted attention to approaches that can be implemented in actual-world systems within a short time [8, 21]. More focus is now on the cell, and pack level approaches. The effectiveness of these solutions primarily depends on the charging strategies used to determine the current density.

2.2.6 Types of Charging Protocols

There are several charging protocols for the Li-Ion batteries: standard protocols, multi-stage constant current protocols, and pulse-charging protocols. A comparison of different charging protocols is presented in Table 2.1.

❖ Standard protocols
Constant current-constant voltage is the most common charging protocol for these batteries. The CC-CV protocol has a constant current charging stage. In the CC phase, the battery voltage increases steadily until it reaches the cut-off value. The CV phase follows, and the constant voltage does not change. Constant voltage starts to change after the current falls to near zero. In the CV phase, there is high capacity utilization because the concentration gradients within electrode particles can easily disperse.

Moreover, there is no need for increasing the voltage as the utilization occurs within the normal range [22]. The charging time during CV increases because of the decreasing current. CC-only charging also has some weaknesses despite having lesser charging time. Therefore, CV-CC has been applied as the standard charging protocol in various applications as it has significantly low charging times and is easy to implement. However, other charging protocols have exhibited fast-charging capabilities, increased efficiencies, and better power retention than the standard CV-CC protocol. However, other charging protocols have exhibited fast-charging capabilities, increased efficiencies, and better power retention than the standard CV-CC protocol (cf. Figure 2.11).

The effects of the CC-CV charging strategy on reversible capacity and anode potential development have been studied. Zhang investigated these

Table 2.1 Comparison of different charging protocols.

Approach	Advantages	Disadvantages	Key elements
CC	Simple implementation	Low capacity	Terminal condition for charging at a constant current rate
CV	Simple implementation. Steady terminal voltage	It is simple to induce the battery's lattice to collapse.	Terminal condition for charging at a constant current rate
CC-CV	High capacity and steady terminal voltage	It is difficult to strike a balance between goals like charging speed, energy loss, and temperature change.	In the CC phase, the current rate is constant. In the CV phase, the voltage is constant.
MCC	Simple implementation and fast charging	It is difficult to strike a balance between goals like charging speed, capacity use, and battery life.	The number of CC stages and their constant current rates

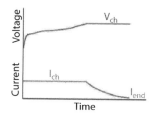

Figure 2.11 Constant current-constant voltage protocol.

effects using graphite cells with reference electrodes in an experiment published in [23]. Temperatures ranging from 2 to 10 degrees Celsius were used to charge the electrodes. C-rates on the electrodes ranged from 0.16C to 1.2C. The results of this experiment revealed a link between increased charging current during the CC stage and increasing time during the CV

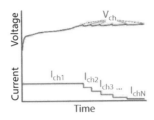

Figure 2.12 Multi-stage constant current (MSCC) charging.

stage. As a result, even though the current was continuously increased, the total charging time did not decrease, and this phenomenon was linked to the CC-CV association.

Additionally, increasing the charging current and decreasing ambient temperature reduced the anode potential. The researchers associated the reduced anode potential as the primary cause of lithium plating, particularly when it gets to a negative value. However, other methods were not experimented to confirm the reduced temperatures and plating occurrence.

❖ Multi-stage constant current protocols
Adjusting current levels during charging can help minimize cell degradation and achieve fast charging. This consequently reduces the charging time [20, 21]. Approaches that involve current level adjustment depend on reducing overall heat generation during the charging process, eliminating or minimizing conditions that enable lithium plating, and reducing mechanical stress that occurs when Lithium-ion diffusion does not happen smoothly. The MCC protocols are among the first types that were designed when the need to achieve fast charging emerged (cf. Figure 2.12). Unlike the CC-CV protocol that consists one of each, MCC protocols have more than one CC stage, followed by a CV phase. When charging batteries that use these protocols, higher current levels are used during the initial stages

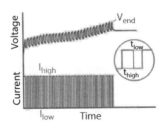

Figure 2.13 Pulse charging (PC).

because the chances of the anode potential becoming negative are low. However, some researchers have done the reverse by increasing the current level during later stages [23]. The latter MCC-CV approach exhibited fast capacity loss than the CC-CV protocols [23]. However, more experimentation is needed because the protocols are yet to be experimented on other cells as their application is limited to pouch cells.

❖ Pulse-charging protocols
These types of charging protocols are also common in literature. The pulse charging protocols involve periodically interrupting the charging current (cf. Figure 2.13). The interruption is achieved using either short rest periods or discharge pulses. The primary purpose for interrupting charging current is to reduce concentration polarization and minimize the risk of the anode potential becoming negative, leading to plating. The strategy also reduces overall mechanical stresses that emerge from uneven insertion and when lithium is extracted from solid particles. Some studies have also revealed that pulse charging can also inhibit dendrite propagation [24]. However, the implementation of this protocol has been associated with a lot of complexity. Moreover, there is increased level of uneven heat generation when using this protocol caused by rebalancing effects as pulses keep changing. There are no studies addressing the implications of these effects on the charging process and cell life.

❖ Literature review on the Fast Charging of Li-Ion Batteries
The ability to achieve fast charge and slow discharge are crucial in the most current utilities of battery systems. For this reason, Battery Management System (BMS) has a significant role in enhancing the performance of batteries to allow for the fastest charging and slowest discharge possible. Today, advancement in BMS systems has allowed many manufacturers to produce electric vehicles [26]. However, the length required to charge their batteries and controlling their energy use is still a major concern to their developers. While in operation, EVs require a high current to facilitate acceleration, which has the drawback of reducing the efficiency of the battery. Achieving fast charging is a significant matter that requires atomic as well as system-level understanding.

Numerous experts have focused on studies on achieving fast charging of lithium-ion batteries [25, 26]. They identify several factors that affect the speed of charging of a lithium-ion battery. The factors include the temperature under which the battery is subjected, the level of current used in charging, the use and maintenance of the battery, and operational factors such as the frequency of use of the battery. Charging at a low temperature significantly boosts the speed of charging [25]. For this reason, numerous researchers have focused on methodologies that utilize low-temperature charging to help in the application of batteries in EVs. Focus has also been placed on controlling the degradation of the lithium plating, which also reduced the battery's charging times and energy-holding capacity. Low-temperature charging is the most widely utilized approach because it provides the potential of developing onboard low-temperature charging methods [26]. Diverse methods are used to facilitate low-temperature charging methods, with other strategies focusing on combining it with managing the lithium plates through onboard systems. Some thermal management methods entail cooling the batteries during charging during normal weather and preheating them in excessively cold weather. The ability to achieve a balance between high-speed charging and operational temperature for the battery system is significant for the optimal performance of EVs.

Monitoring the degradation of the lithium plating is a difficult task that has not been achieved through onboard methods today. As a result, EVs have an extra burden of requiring examination of the battery system from the service station [26]. Consequently, achieving fast charging of batteries for EVs has been a major concern for experts. Therefore, approaches for detecting the degradation of the lithium plating in batteries or the breaking of the plating have been based on examining the voltage plateau [26]. However, there are significant challenges in distinguishing between the stripping of lithium from other factors that also induce a voltage plateau.

Similarly, detecting the degradation of the plating where a plateau has not occurred is another challenge, owing to the fact that the plateau does not always follow a deterioration of the lithium plating [26]. Nevertheless, many experts suggest key approaches of causing fast charging of lithium batteries. They include searching for a possible material to replace the lithium plating, exploring new modeling methods that could offer better charging speeds, and expanding studies on low-temperature charging, which could offer more benefits to the system.

Experts continue to search for other electrode materials for batteries that could address the plating degradation phenomenon of lithium batteries [27]. While they suggest quite a number, there is a need to examine them to ensure that they can match the stability, cost, and ease of manufacture of the lithium plating. Graphite is the primary material currently used for lithium plating. It is likely to continue to be the most preferred material despite its observed drawbacks in slowing down the charging process. Significantly, it is easy to manufacture and does not involve high costs, which makes it a viable option so far.

Exploring additional models of enhancing the charging speed of batteries is another possibility among researchers. Specifically, the subject of study regarding lithium batteries is broad and unexhausted, giving opportunities for further studies. Existing models have substantial limitations. Additional methods could be established of examining the internal conditions of batteries and providing the necessary changes to facilitate optimal operations. Only a few physics-informed methods have the capacity to perform such examinations, but they still have a limited range of use and cannot be utilized in physically abusive conditions that EVs are bound to encounter. Therefore, conducting studies on fast-charge models could provide new models that could offer viable options.

A study by Xing et al. also affirms the limitations of numerous models of implementing current fast-charge capabilities in battery management systems [28]. The author recognizes the need for conducting studies and evaluations of the performances of the system in all possible temperature and weather conditions. Significantly, the batteries are bound to undergo numerous environmental conditions with varying degrees of several factors. Notably, the range of possible temperature experiences are wide as EVs may be required to operate in high-temperature situations with low humidity to excessively low ones with high humidity [28]. Furthermore, the possibilities of having various combinations of temperature, humidity, and other astute environmental conditions are numerous, making the study significantly complicated. Fortunately, the presence of numerous possible environmental conditions provides numerous possibilities for developing advanced systems that offer the best solutions to the limitations currently experienced on the systems.

Xing et al. focus their study on measuring the various "states" of lithium batteries, including the state of charge and state of health. The authors review numerous studies on evaluating the state of the batteries and reveal significant challenges in findings that differed in respect to specific environmental conditions [28]. Similar to other researchers, they recognize that batteries for use in EVs and HEVs have to offer fast charge by having a high operational current and voltage. As a result, batteries for EVs and HEVs are significantly more complicated compared to those used in other diverse types of devices and equipment. According to the researchers, a layered decentralized topology is the most appropriate for fast charge and long-term use batteries [28].

Additionally, they recognize the need to implement systems that could offer real-time evaluations of the parameters experienced by the battery. These include their thermal conditions in each cell, the discharge strategy, the cell equalization, and charge times. Evaluating the state of the battery provides essential information about the temperature condition that the battery is subjected. Additionally, it provides an array of other useful factors, such as the operations of the charging and discharging mechanisms that facilitate cell balancing. One of the primary challenges is offering onboard systems to conduct the evaluations because the systems will also be subjected to the diverse environmental conditions that the battery is exposed [28]. Consequently, their calibration may be altered, leading to erroneous assessments. Considering that the continual assessment is essential to control the entire system, the failure of the onboard system could cause significant damage to the battery or explosion of the system that could have disastrous effects on the EV or its occupants.

The authors also point out the complications of developing a model for the battery system. They argue that modeling for state of charge encompasses applying various charging networks that have diverse material characteristics and numerous possible results based on the diverse requirements for accuracy [28]. For this reason, current models provide insufficient conditions to offer substantive rate of charge evaluations in diverse environmental conditions. Consequently, the reliability of battery management systems in diverse environmental conditions is difficult, which creates significant challenges for BMS systems for EVs and HEVs.

To achieve cell balancing in EVs and HEVs, a common approach is using parallel wiring systems that lowers the effective load of the system. As a result, the parallel systems have lower current but relatively higher voltages [29]. To balance out the need for a high voltage, some systems in the electronic model are wired in series, which enhances their voltage without necessarily affecting the current. However, the two setups create complications because weaknesses or defects in one of the cells in a series connection style affect the rest of the cells [28]. For this reason, overcharging in one of the cells could affect the state of charge of other cells, which deteriorates the state of the entire system. The cell balancing functionality works by keeping the state of charge of all cells contained within the entire battery system as close as possible to minimize the effect of any defect in one of them.

2.2.7 Li-Ion Battery Degradation

Li-Ion battery degradation is caused by physical and chemical mechanisms that impact the battery's components, including the electrolyte, the electrodes, the separator, and the current collectors [30]. The capacity

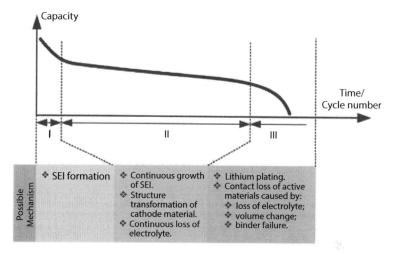

Figure 2.14 The main stages in the capacity fade vs. time.

of a deteriorated Li-Ion battery is lost, as is the ability to retain a charge. Resistance increase is defined as a decrease in the rate at which electrical energy may be received or released by the battery as a result of such deterioration.

The aging characteristics of most batteries are nonlinear [31]. Figure 2.14 depicts the three major stages of capacity fading as a function of time. The battery's capacity rapidly decreases in the first stage owing to SEI development. Because of many side reactions that occur inside the battery, battery performance gradually deteriorates in the second stage. The rate of capacity loss accelerates in the third stage, and cell breakdown happens swiftly. The following factors contributed to the failure:

1. Lithium-ion inventory loss due to lithium deposition [31, 32]
2. Active material loss due to the degradation of electrolyte [33]
3. Failure of the binder and volume change [34].

2.2.8 Factors that Cause Battery Degradation

❖ Calendar life (Time)

Calendar life is shelf life. It is the time before a battery, whether in inactive use or inactive, becomes unusable. Balagopal *et al.* [36] state that calendar aging has an important impact on a battery's performance. Over time, the capacity of the battery declines, and it causes degradation of the battery.

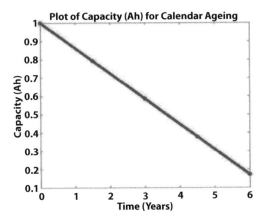

Figure 2.15 Battery capacity vs. the time curve.

Degradation occurs because during manufacturing a thin SEI is formed to prevent active lithium ions and the electrode material from being exposed to the electrolyte. With battery use, degradation occurs because as the SEI grows, it reacts with electrode material and Li-Ion. Figure 2.15, adapted from [36], shows the impact of calendar life on battery capacity.

❖ **Temperature effects**
Temperature plays an important role in Li-Ion battery degradation because it affects chemical reactions that occur inside the battery. The Arrhenius equation, shown below, describes the relationship between the rate of chemical reactions and the reaction temperature. The equation clearly shows that the reaction rate inside the battery increases exponentially as the temperature rises.

$$k = Ae^{\frac{-E_A}{RT}} \quad (2.1)$$

Where k is the rate constant, A is the frequency factor, R is the gas constant, E_A is the activation energy, and T represents the temperature.

Table 2.2, adapted from information obtained in [30] and [37], shows the impact of high and low temperatures on Li-Ion batteries.

❖ **High SoC & Low SoC**
The ratio of the available capacity in the battery to the maximum potential charge that may be held in a battery is known as the state of charge

Table 2.2 The effect of extreme temperatures on Li-Ion batteries.

	Low temperature	High temperature
Impact on battery	Li+ plating	The development of the solid electrolyte interphase is accelerated (SEI)
		Decomposition of electrolytes
		Decomposition of the solid electrolyte interphase (SEI)
		Decomposition of binders
		Dissolution of transition metals and development of dendrites

(SoC) [38]. The cell is considered to be completely charged if its state of charge is 100%. The cell is entirely drained when the SOC reaches 0%.

The capacity fading processes, according to [39], are extremely complicated and little understood.

Changes that occur in a low state of charge and changes that occur in a high state of charge can both be categorized as mechanisms. A low charge induces structural changes in the electrolyte, such as a breakdown process and Mn2+ dissolution. Thermodynamic instability and electrolyte oxidation are caused by a high state of charge.

❖ **Charge-discharge cycles**
The capacity of the battery is reduced as it is charged and discharged. Excess Li-Ions are deposited on the electrode's surface in a Li metal layer termed lithium plating when current is driven into the battery during the charging process. Lithium plating begins to form in a tiny area of the anode near the

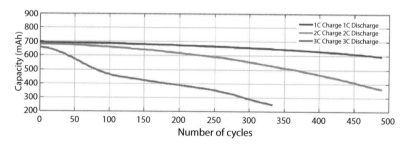

Figure 2.16 Number of cycles vs. capacity at different rates.

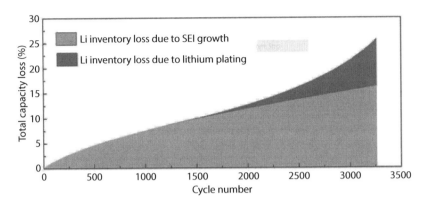

Figure 2.17 SEI build-up with the increasing numbers of cycles.

separator after a number of cycles, according to Yang *et al.* [32]. The chemical conversion process of the active material cannot meet the output of the battery current after too many discharges. Figure 2.16 [16] displays the number of cycles vs. capacity at various charge and discharge rates. As the number of cycles rises, battery capacity declines. Figure 2.17 [13] indicates that as the number of cycles grows, so does lithium plating and the SEI.

2.2.9 Degradation Mechanism of the Li-Ion Battery

The degradation mechanism of the Li-Ion battery is a chain of events that causes degradation effects. The following are the common degradation mechanisms of the Li-Ion battery [30]:

❖ **Loss of Li inventory**
Lithium ions are eaten by parasitic processes, which combine electrochemical and chemical reactions [30, 40], resulting in a loss of Li inventory. SEI growth, decomposition reactions, and lithium plating are examples of parasitic processes. Li ions are no longer accessible for cycling between the positive and negative electrodes when the Li inventory is depleted.

❖ **Loss of active anode material**
When the active mass of the anode is not available for the insertion of lithium, the active material of the anode is lost [30]. This can happen in any of the following scenarios:

1. Particle shattering
2. A break in electrical contact
3. The use of resistive surface layers to restrict active areas.

❖ **Loss of active cathode material**
When the cathode is no longer available for the insertion of lithium, there is a loss of active cathode material [30]. This might happen in any of the following scenarios:

1. Disorganization of the structure
2. Cracking of particles
3. A break in the electrical connection.

❖ **Increased electrical resistance**
The internal resistance of a Li-Ion cell can be increased by delamination of the current collector from the electrode owing to gas development in the electrolyte [41, 42]. Decomposed electrolyte molecules combine with lithium to produce insoluble compounds, which become part of the SEI and raise the cell's inner resistance [43]. Other degradation events, including as cathode deterioration and SEI film formation on the anode, are accelerated by the higher temperatures, resulting in capacity fading and a rise in internal resistance [44].

Figure 2.18, adapted from [30], provides an overview of the main degradation mechanisms in Li-Ion battery. The degradation mechanisms are very complex because the side reaction inside the cell is related to the materials [35], and the degradation mechanism is caused by different degradation mechanisms and at different rates. Due to the complexity of the degradation mechanism, physics-based models focus only on the dominant mechanisms of degradation, such as the solid electrolyte interphase (SEI) [45, 46]. Tables 2.3, 2.4, and 2.5 provide an overview of the factors,

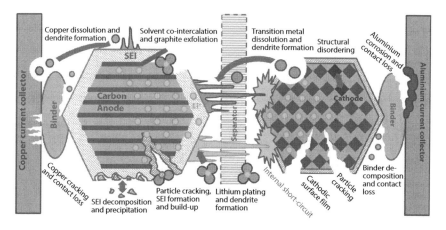

Figure 2.18 An overview of the main degradation mechanisms in the Li-Ion battery.

Table 2.3 The root causes and consequences of Li-Ion degradation.

Root cause	Consequences							
Time	SEI growth							
High Temperature	SEI growth	Electrolyte decomposition	SEI decomposition	Binder decomposition	Transition metal dissolution/ Dendrite Formation			
High SoC		Electrolyte decomposition	SEI decomposition	Binder decomposition	Transition metal dissolution/ Dendrite Formation	Graphite Exfoliation	Li+ plating	
Low SoC								Current collector Dissolution
High Current	SEI growth		SEI decomposition			Graphite Exfoliation		Loss of electrical contact
Low Temperature							Li+ plating	Loss of electrical contact	Structural disordering	Particle Cracking & Island formation

Table 2.4 The causes of the degradation mechanism in the Li-Ion battery.

Degradation mechanism	Caused by											
	SEI growth	Electrolyte decomposition	SEI decomposition		Binder Decomposition	Transition metal dissolution/ Dendrite Formation	Graphite Exfoliation	Li+ plating	Current collector Dissolution	Loss of electrical contact	Structural disordering	Particle Cracking & Island formation
Loss of cycable Li	SEI growth	Electrolyte decomposition	SEI decomposition									
Loss of Active Anode Material							Graphite Exfoliation					Particle Cracking & Island formation
Loss of Active cathode Material					Binder Decomposition	Transition metal dissolution/ Dendrite Formation			Current collector Dissolution	Loss of electrical contact	Structural disordering	Particle Cracking & Island formation
Increased Electrical Resistance					Binder Decomposition				Current collector Dissolution	Loss of electrical contact	Structural disordering	Particle Cracking & Island formation

Table 2.5 The degradation mechanism and its effect on the Li-Ion battery.

Effect on battery	Degradation mechanism				
Capacity Loss	Loss of Cyleable Li	Loss of active anode material	Loss of Active cathode material		
Resistance Rise		Loss of active anode material	Loss of Active cathode material	Reduced Kinetics	Increased Electrical Resistance

causes, and effects, respectively, of degradation mechanisms and the associated degradation.

2.2.10 Electrode Degradation in Lithium-Ion Batteries

Electrodes play a significant role in the degradation of Lithium-Ion batteries. Understanding electrodes degradation and the behavior is very significant to allow us to design the best fast-charging protocols. Therefore, in the next section, we will review electrode degradation in Lithium-Ion batteries.

According to Mei [47], the physical properties of electrodes have a significant impact on the rate of heat generation and energy density of batteries. Thicker electrodes, according to Zhao [48], can increase the fraction of active materials and hence enhance energy. However, increasing electrode thickness can have a detrimental impact on Li-Ion battery thermal and electrochemical performance. He also shows that when batteries with a thicker electrode discharge at the same rate, they have more intense and uneven temperature reactions across the electrochemical cell. This can result in active components being depleted and the battery's capacity diminishing faster. Furthermore, a Li-Ion battery with a thicker electrode is more susceptible to a higher discharge rate, he finds. This is due to the formation of ohmic heat, which causes the Li-Ion battery's health to deteriorate.

Figure 2.19 depicts the weight % breakdown of a commercial Li-Ion battery per component. The electrodes are definitely the most important material components of the Li-Ion battery, as seen in the diagram. Because electrode deterioration has such a substantial influence on Li-Ion battery

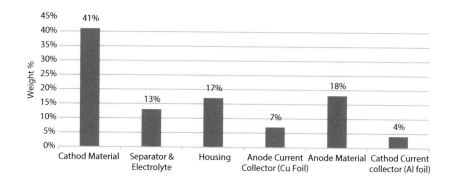

Figure 2.19 The breakdown of the weight percentage by the components of a commercial Li-Ion battery.

performance, and because electrodes are the biggest and heaviest material components of the Li-Ion battery, electrode degradation has a considerable impact on Lithium-Ion battery lives.

❖ Degradation mechanisms on the anode materials
The lithium-ion will be associated with the electrolyte breakdown while charging at a fast rate. The SEI is a passivation protective layer formed on the surface of the anode electrode as a result of this breakdown (solid electrolyte interface). After charging and discharging, the volume of the anode material varies by 10% [35]. This is due to lithium ion intercalation and deintercalation (cf. Figure 2.20). This shift in anode volume can cause the SEI to break, resulting in a lithiated graphite and electrolyte reaction that consumes the Lithium-ion and electrolyte. As a result, the SEI coating on the graphite anode's surface thickens.

❖ Degradation mechanisms on the cathode materials
The deterioration of the cathode in LiFePO4 cathode material is caused by the breakdown of the binder. Another source of cathode deterioration, structural disordering, is current collector corrosion driven by parasitic processes [50, 51]. Insoluble species that migrate through the separator and dissolute in the electrolyte may cause side reactions at the cathode in some situations [39, 63]. The oxidation of electrolyte components and the development of surface films are two further degradation mechanisms [39] (cf. Figure 2.21). Surface electrolyte reactions [39] and the consequent creation of fissures in particles cause gas evolution.

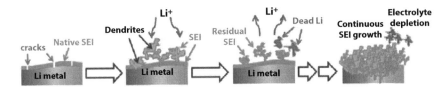

Figure 2.20 Degradation mechanisms on the anode material.

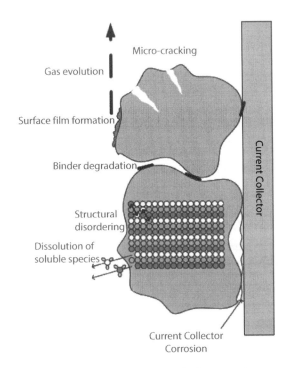

Figure 2.21 Degradation mechanisms that affect cathode materials.

2.2.11 The Battery Management System

A battery management system is a crucial strategy for extending the battery's life and managing quick charging. The Battery Management System (BMS) has several functions, including protecting the operator's safety, detecting unsafe operating conditions, protecting battery cells from damage in failure, and informing the application controller how to make the best use of the pack right now (e.g., power limits), and controlling the charger.

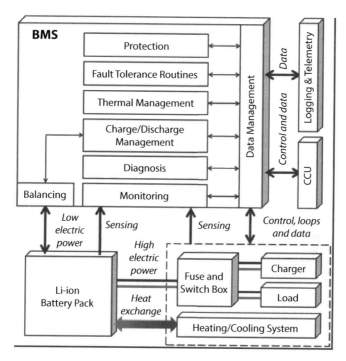

Figure 2.22 Typical battery management system.

BMS manages the output of a battery and provides notifications on the condition of the battery. The BMS manages the battery's charging and discharging processes and provides vital precautions to protect rechargeable batteries from damage [58]. Different types of BMS are used in different devices using rechargeable batteries. Battery management systems are commonly found in cars, data centers, smartphones, and mp3 players.

The BMS monitors and regulates the main power voltage, charging and discharging levels, temperatures of the batteries, and how healthy the battery is [53]. Also, the systems check and control the battery voltage and the coolant temperature [54]. Figure 2.22 shows a sample block diagram of a typical BMS system.

❖ How BMS Can Help Increase the Life and Reliability of Battery

The lifetime of a battery depends on several aspects ranging from its internal battery parameters to external parameters. The internal parameters are influenced by the manufacturers, while the external parameters are influenced by the users [55]. The external aspects can have a major impact on the lifetime of a battery. Through the use of the BMS, it is possible to control the external parameters to increase its life. Considerations of the demands and the battery life objectives help establish the correct architecture to create a BMS and charging strategy that optimizes the life of the battery [56].

BMS is needed in providing safety and proper battery performance. Improper battery conditions such as over-voltage and over-charging cause rapid aging and can lead to an explosion. BMS includes sensors that estimate vital parameters such as temperature, voltage, and current. After determining the thermal and electric behaviors of batteries, it is important to optimize the charging and discharge algorithms to provide efficiency and enhance the battery lifespan [57]. The BMS alarm and safety control module helps eliminate any abnormalities that might make the battery less reliable or shorten its lifespan [57]. The parameters of a BMS depend on the battery operation conditions. Therefore, it is significant to collect data on the battery's behavior during the charging or discharging processes. The BMS helps to increase the life and reliability of a battery by stopping charging and discharging current if it establishes that it is necessary.

Batteries are used in electric vehicles, and BMS plays a significant role in enhancing their reliability. The battery systems of electric vehicles are made of several battery cells packed together. The systems have a high voltage rating and current, and any slight mismanagement action could lead to a disaster. In this regard, BMS plays a significant role in the safe operation of high voltage batteries used in electric vehicles. BMS monitors the state of the batteries and prevents overcharging and discharging that might reduce the capacity and lifespan of the batteries [58]. A BMS checks the voltage and shuts down or sends an alarm when the required voltage is attained. Also, battery management can relay information about the condition of the battery to the power management systems. This helps in correcting any fault before it becomes serious. Also, BMS controls the temperatures of the battery cells and their health, ensuring that the battery is safe and reliable.

❖ **Thermal Management**
Preheating in cold conditions. Maintaining low temperatures for quick-charging Li-Ion cells, as previously stated, has proven to be a considerable difficulty. The literature seeking to address this topic has considerable gaps. Some research, on the other hand, has sought to devise charging procedures that include diverse preheating techniques. This section discusses techniques that can result in fast heating across the cell. The emphasis is on this since high speed is an important feature of any preheating technique that may be used in conjunction with rapid charging. Internal preheating is favored because it ensures that less heat is wasted to the environment, making it more efficient. Li-Ion cells that are particularly built to permit rapid preheating have

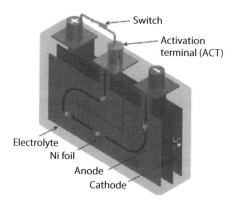

Figure 2.23 Schematic of Li-Ion cell structure with embedded Ni foil.

been used to address fast charging without the risk of excessive temperature effects [59]. As seen in Figure 2.23 [60], this is accomplished by sandwiching electrically insulated nickel foils between two single-sided anodes. Using conventional direct current, the technique assures that the cell can raise the temperature on its own when needed [59]. In most cases, an activation switch is utilized to control the amount of current that passes through the foils [59]. Turning on the activation switch, for example, sends direct current to the foils, resulting in fast heat buildup [59].

Turning it off, on the other hand, permits current to flow between the electrodes while the nickel foil stays separated. Internal heating techniques, according to most studies, boost efficiency and allow for uniform temperature distribution across the cell. However, little study has been done on the effects of internal preheating when utilized in conjunction with rapid charging. Experimenting with internal temperature is difficult. As a result, models that can analyze preheating when alternating current is employed are required. Preheating using nickel foil appears promising; nevertheless, it can only be utilized in non-standard cells, which adds weight and reliability concerns. Furthermore, the concept was very recently proposed, and only a limited amount of study has been conducted on its use. As a result, substantial and additional study into its performance, as well as an evaluation to establish whether the technology is cost-effective, are required before it can be completely commercialized.

2.2.12 Battery Technology Gap Assessment for Fast-Charging

As discussed in the previous sections, electrodes will impact fast charging and cell performance. In this section, we will discuss the other factors and components of Li-Ion batteries which will impact fast-charging.

❖ Electrolyte degradation

Electrolytes are among the cell performance determinants. The electrolyte can affect cell performance and electrode behavior. For example, the electrolyte can potentially change the structural aspects of graphite electrodes [61]. Reports have shown large and irreversible capacity loss of graphite electrodes in batteries with cells that contain propylene carbonate electrolytes. These reports hypothesized that reduction products in these cells lack the capacity to coat graphite as required. Therefore, propylene gas formed after reduction remains trapped within the electrode.

As a result, there is a build-up of high pressure, which causes exfoliation.

There is currently inadequate data on this influence on electrolyte deterioration. Lithium plating and excessive heat production are thought to decrease electrolyte conductivity, according to "Electrolyte design for fast-charging Li-Ion batteries". The addition of low-viscosity solvents to standard carbonate electrolytes to boost ionic conductivity is one of the most promising approaches for improving electrolytes for quick charging. Low-MW esters like Methyl acetate and Ethyl acetate, as well as nitriles, are the best options.

❖ Charging protocol

Several reports have been published about the effects of the mode of charging or protocol used has on Li-Ion cells performance and life. [61] identify rapid degradation when various protocols are used. For instance, one observation showed the various methods discussed earlier cause different lithium deposition levels. Like most researchers, [61] emphasize that field implementation of the charging protocols will require the development of complex algorithms, which are yet to be studied.

❖ Usage

Customer usage is a core factor in battery development. Manufacturers must consider consumer usage patterns during design and optimization processes. The customer usage pattern is considered because it affects battery life, performance, and safety; this makes the conventional CC-CV charging protocol unsuitable for extreme fast charging (XFC) because it is prone to forming a lithium plating that will compromise performance and safety. The suggested fast-charging protocols and other newer protocols should be explored. Electric vehicle users are likely to arrive at recharging stations with empty to almost full battery packs. This depends on extreme fast charging (XFC) availability, convenience, and pricing [61]. Research shows that currently, auto manufacturers assume that EV owners charge their cars at home, especially after work. Data collected from drivers who adopted EVs early has proved this assumption right as most EV users charge at home and work during their break hours. However, it is difficult to determine what the duty will look like when a larger part of the population switches to electric vehicles as more countries are encouraging the use of renewable energy resources in the transport sector. From the battery perspective, manufacturers should focus on developing batteries that correspond to aggressive use by reducing the overall charging hours and improving performance and safety.

❖ **Fast-charging effects on thermal runaway characteristics**
Besides degradation, fast charging has raised some safety concerns. Lithium plating and temperature rises that occur during fast charging pose potential risks. For instance, research has shown significant changes in batteries' thermal runaway behavior after fast charging [59]. Accelerating Rate Calorimetry (ARC) tests on fast-charged high-energy batteries revealed that the self-heating temperature reduced significantly when cells were subjected to fast charging than it dropped in fresh cells. However, these effects are not permanent since, with adequate rest time, they are reversible. Besides this, overcharge-induced thermal runaway is another safety concern for fast-charging batteries. There is a possibility of having some cells in a battery pack overcharging after fast-charging. This phenomenon is often caused by inconsistencies among the battery cells.

Two design models have been proposed to increase safety in fast-charged batteries by protecting them from overcharging. One is by using an electrolyte with a higher oxidation potential. For instance, raising the oxidation potential from 4.4V to 4.7V will increase electrolyte stability. This can be achieved by including additional functional additives. Functional additives are chemical species that can be oxidized at a higher potential than the conventional safe cathode potential. Considering the unique Li-Ion cell components, there is a need for more research to determine the suitable redox shuttle species that can be used in these batteries without possible risks and effects like cell-life reduction. Secondly, increasing the temperature at which the battery thermal runaway occurs can help reduce risk levels in fast-charged batteries. The temperature can be increased by increasing the battery pressure relief design to the optimum.

2.2.13 Developmental Needs

Based on Li-Ion battery gaps assessment and challenges around this technology, some developmental needs have been proposed to enable fast charging. Changes in modeling are among the needs that [61] proposed.

Extreme fast charging (XFC) requires advanced models; therefore, the current ones need to be updated. Advanced models will be incorporated with fast-charging protocols [62, 63]. Model development should also include fast-charging constraints and other design requirements to facilitate cost estimation for the entire process [64, 65].

Additionally, the updated models should include the identified temperature, and current effects on the life and performance of fast-charging-enabled battery packs. Aging and failure due to degradation are the most pertinent fast-charging mechanisms. Therefore, manufacturers need to identify and include them in the advanced models.

2.3 Materials and Methods

The research utilized secondary resources. Existing literature about fast-charging protocols for Li-Ion batteries was the primary resource for the study. Various articles assessed the electric vehicle battery technology and other factors around the fast-charging protocols. The study also relied on experimental data. Researchers assessing these technologies have used experiments to evaluate the effectiveness of different fast-charging protocols [20, 21]. The experiments use cells with different rating capacities and under varying current profiles. The experimental conditions in the reviewed literature are changed to assess the thermal behaviors of batteries when different protocols are changed. The literature also provides different models for various charging protocols. For example, in their work, [21] presented a model to calculate the charging current profile as a function of the charging stage. When designing a fast-charging protocol for Li-Ion batteries, the model revealed the significance of taking cost functions into account. The researchers recommended a multi-stage pricing methodology based on their findings [21]. When evaluated in the model, the technique was shown to be successful in lowering the capacity fade ratio.

All of the cells used in the experiment were cycled using different charging techniques, according to one of the studies published in the literature. They were, however, fired in the same manner.

The ten-minute charging methods were used to charge the cells. [66] was charged in the range of 0% to 80% SOC. After that, they were granted a five-second break. Following that, all cells were charged with an 80% to 100% SOC to the manufacturer's specified optimal voltage and a C/20 cutoff. After another five seconds of rest, the cells were discharged with a CC-CV discharge of 4C to 2.0V and a C/20 cutoff [66]. After the discharge, the cells were allowed to rest for another five seconds before starting the charging process [66]. The lowest and upper cutoff voltages were 2.0V and 3.6V, respectively. Cycle life refers to the total number of cycles until the discharge capacity falls below 80% of the nominal capacity in all experiments conducted by previous researchers.

Using the charging and discharging procedure described above, the cell capacity of each cell was measured after every 10 cycles. Q [66], a dimensionless capacity, was introduced by the researchers. Q was the ratio of the old cell's capacity to the new cell's capacity. The plotting of the dimensionless values revealed a considerable increase in the aging rate. A rise in the cut-off voltage is related with rapid acceleration. Some phenomena were observed during the studies, such as a high concentration of surface ion, which limited the charging current. Concentration is advantageous because it avoids capacity deterioration.

A two-stage constant current rapid-charging methodology is evaluated in a second experiment published in the literature. The batteries are charged at 80% of the C-rate for half an hour in this experiment [20]. The goal of the study was to examine the thermal behavior of various charging current patterns [20]. The electrochemical-thermal-coupled model is used to analyze the thermal behavior and charging efficiency of this approach. For a cylindrical Li-Ion phosphate battery, the model was constructed using charge and energy conservation concepts [20]. The charging current profiles for the two-stage constant current protocol achieve higher charging efficiency and lower temperature than the quickest charging protocols published in the literature, indicating that the experiment yielded encouraging results.

Another research, [67] called "Reliability and Failure Analysis of Lithium-Ion Batteries for Electronic Systems." The researchers in [67] used two commercial batteries designed for application in portable electronic systems were used for testing the impact of increasing charging rate for batteries with the same material but with different physical parameters. Table 2.6 below shows the physical parameters for each battery.

Table 2.6 The physical parameters for each battery.

	B1	B2
Nominal Capacity (Ah)	1.1 Ah	1.35 Ah
Length (mm)	50	50
Width * Thickness (mm)	33.8*5.4	33.8*6.7
Weight (g)	22	28

In this study [67], it was found that a battery with a thicker electrode will be very sensitive to the increase of the charging rate; this is due to the ohmic heat generated from the battery, which will lead to the deterioration of the battery.

2.4 Discussion

Battery technology is critical in electric car commercialization which makes a collective effort to address the climate change crisis resulting from pollution in the transport sector. Literature shows that electric vehicle manufacturers have widely adopted lithium-ion batteries. The Li-Ion battery technology has been widely reviewed, with more studies focusing on fast charging and the key aspects that limit it in these batteries. Various fast-charging protocols for Li-Ion batteries have been proposed, with some like multi-stage constant current MSC proving to be effective in enhancing fast charging. However, the literature identifies various limitations to fast charging, lithium plating being the primary challenge to the technology.

A positive electrode constructed of metal oxide is found in a conventional Li-Ion cell. Organic carbonates with lithium-bearing salt and graphite rod are also included in the cell. The negative electrode is a graphite rod, while the electrolyte is an organic carbonate mixture [61]. There is an ion movement during charging, with Li-Ions moving from the metal oxide to the negative electrode via the electrolyte as the medium. During discharge, the ions flow in the opposite direction. Electrode particles are covered with other operation products as a result of the reaction. The sold interface layer (SEI) is formed by the coating [61]. The SEI components differ depending on the electrode, with the negative electrode having a higher value.

Li-Ion batteries' ability to maintain high power densities makes them ideal for usage in electric cars. Recharging Li-Ion batteries takes longer than fuelling an internal combustion engine car, notwithstanding its efficacy. Consumer adoption of electric vehicles has been hampered by the passage of time. Furthermore, due of the range anxiety issue connected with electric vehicles, drivers are hesitant to adopt them. The capacity of manufacturers to guarantee that cars have a shorter recharge time that is equivalent to that of internal combustion engine (ICE) vehicles is critical to achieving efficient customer adoption. Fast-charging techniques for Li-Ion batteries are used to reduce refueling time. Additionally, unlike with carbon-based fuels, battery recharge should not be confined to empty to fully charged batteries. Consumers should be able to charge their automobile batteries to any capacity they choose. The electrolyte-negative electrode contact is reached as Li-Ions are transferred from the positive electrode to the negative electrode during charging. Lithium metal is likely to plate on the negative electrode when the transportation process is quicker than the intercalation rate, according to studies [61]. Lithium plating has been shown to increase at low temperatures and with high current density [61]. Even at temperatures as low as -20 degrees Celsius, lithium plating can occur at a C-rate of 6.

System flaws have been documented to produce plating, in addition to higher current density and low temperatures. For example, plating is likely to occur if pore closure occurs in the separator or if there are large currents present throughout the operation. If the overpotential is larger than the equilibrium potential, plating is likely to occur [68]. The lithium quantity deposited on the negative electrode depends on the electrode's capacity loading. [61] show that in some cases, the plating is removed during the discharge cycle. However, discharge is unreliable for eliminating plating because even slow discharges have been seen not to remove significant amounts of plating from the negative electrode [16]. The findings suggest that the connection between lithium deposits and the graphite rode is not electronic. Under severe circumstances, the deposits affect cell performance and cell life. Figure 2.24 below is adapted from [61], and it shows how lithium deposition increase with increasing the charge rate.

In the extant literature, there are several ways for detecting lithium plating. For example, [64] and [69] provide non-destructive ways for detecting lithium deposits on graphite rods. Volumetric measurements to identify any anomalous cell volume variations, voltage monitoring to detect high-voltage plateaus, and high-precision coulometry to detect cell efficiency changes are examples of these approaches. Furthermore, continuous

Figure 2.24 Lithium deposition increase with increasing the charge rate from left to right.

current charging guarantees that the rate of lithium delivery to the negative electrode remains constant. Placing on the electrode surface is minimized by keeping the delivery rate equal to or less than the intercalation rate. Despite maintaining such conditions, however, lithium plating avoidance is not always assured since other causes, such as a rise in the local chemical potential, might produce plating.

The study literature also includes studies on the effect of capacity loading and charge rate on lithium plating [70]. Lithium plating as a function of C-rate and capacity loading is shown in [16]. Plates were predicted to appear in cells with the largest capacity fade. This was proven by the study, which found that these cells had the most lithium deposits [16]. Despite the fact that the cells were discharged at a moderate rate, lithium deposits persisted at the electrode, contrary to predictions. This shows that when lithium deposits are plated, they become electrically insulated from the negative electrode. Furthermore, the findings imply that theorized reconditioning of the battery pack by gradual discharging cannot return accumulated lithium on the graphite electrode surface to the positive electrode.

With fast charging, the lithium deposition rate increases, increasing degradation, consequently reducing the cell life. The rate at which the cell degrades and its overall performance depends on temperature. Besides the temperature at which a cell operates, other factors, such as the nature of active material and graphite rode design, also determine degradation level. Graphite is the common material used in lithium-ion battery negative electrodes. Plating occurs easily on graphite electrodes because of the local chemical and electrochemical around the material. However, other materials are explored for their effectiveness in use for the negative electrode in lithium-ion cells. These materials have a higher potential than graphite. Higher potentials are beneficial as they limit the conditions needed for lithium plating. Lithium metal is also suggested as a graphite alternative for the negative electrode. However, as evident from the studies, lithium metal is associated with dendrite formation, reducing the life cycle and increasing safety concerns [8]. As a result, before employing lithium metal as a negative electrode, these problems must be resolved. Lithium metal, according to some studies, cannot be utilized in applications that need a lot of power [61]. Lithium was discovered to be very reactive when combined with various electrode components. In addition, the metal generates a more complicated SEI layer than other materials. Dendrites still occur when lithium metal is utilized, despite the presence of the SEI layer. The surface morphology of electrodes made of lithium metal was also shown to be sensitive to current density [61]. The surface of the lithium metal altered from smooth to dendritic in most studies, making the surface rough.

When lithium metal is exposed to rapid-charging conditions, dendrite growth is accelerated, rendering the material unsuitable for the application.

(Li4Ti5O12) LTO is considered to be the appropriate graphite electrode alternative. LTO (Li4Ti5O12) can support fast charging as it has the required electrode kinetics. Several literature reports show that (Li4Ti5O12) LTO nano-particles can continuously be charged at C-rates as high as 10C without any additives [61]. Additionally, doping the material increased its electrochemical performance. [20, 21] and [59] further reported that sodium-bearing phases have a higher rater performance than (Li4Ti5O12) LTO. Like (Li4Ti5O12) LTO, doping the phases increase their superiority by improving electronic conductivity and ionic diffusivity. Silicon is another material that could be used as a negative electrode; however, its response to fast-charging conditions has not been assessed in the resources used for this study. Therefore, the viability of silicon electrodes and their stability in fast-charging applications is not currently known.

The open literature has yet to assess the impact of fast-charging conditions on the positive electrode. However, some reports show that metal oxides, which are the primary positive electrode materials, have low stability for cycling [61]. Charging involves the diffusion of ions in and out of the host material. This leads to changes in volume and concentration gradient around the positive electrode. Diffusion-induced stress can potentially cause failure, especially when the C-rate is high. Stress levels are likely to increase during fast charging, causing degradation when temperatures are non-uniform. However, there is a need for more research assessing the effects on the positive electrode as there are significant knowledge gaps on the phenomenon.

Electrode design is another factor identified to influence plating. [16] shows that electrolyte transportation and polarization can potentially cause lithium plating on the negative electrode during extreme fast charging. It was proposed that avoiding high charge current densities can limit plating. Charge currents above 4 mA should only be applied after taking necessary precautions. Similarly, as noted in several studies, Li-Ion battery power depends on temperature [18, 20, 21, 59, 61]. The total cell impedance changes significantly over a wide range of temperature changes [52]. During fast charging, resistance is high, causing heating that consequently increases temperatures within the battery cells. High temperatures reduce battery resistance as kinetics become faster. On the other hand, electronic resistance in the electrode current collectors increases as temperatures rise. This allows some power resulting from faster kinetics to be released. It is essential to avoid reaching the upper-temperature limit set by the battery manufacturer during fast charging. This is recommended because if at any point the temperature of the lithium-ion cell when it is fully charged exceeds the manufacturer's pre-determined point, it could lead to a thermal runaway, which is a severe safety concern. Despite some systems having set points that are considered low, they should not be exceeded. Additionally, high temperatures in the electrolyte can cause slat decomposition, shortening the cell life.

Electrolyte degradation is another key aspect of fast-charging protocols discussed in many reviews. The electrolyte has significant impacts on cell and electrode behavior. One of the evident effects is its potential to change the structure of the graphite electrode. For instance, a significant capacity loss was experienced on negative graphite electrodes used in cells containing propylene-based electrolytes. It was hypothesized that reduction products did not coat graphite well. This led to the formation of propylene gas that was consequently trapped in electrode surface crevices, leading to excessive pressure buildup. High pressure caused exfoliation. Currently, there is insufficient information about fast charging on degradation. There are speculations that high heat generation during fast charging and lithium plating affects electrolyte properties. Consequently, further research is needed to determine the exact impact of fast charging on electrode performance.

Because of the necessity to increase rapid charging efficiency in Li-Ion batteries used in electric vehicles, academics have focused on fast-charging procedures. Based on the findings, it is clear that a battery's charging technique impacts its performance decline. In [59], the authors recognize that the CCCV charging protocol has been widely adopted in many batteries and electric vehicle industries. The high adoption is primarily because this charging protocol is simple and less costly than others like the multi-step charging protocol. The CCCV, on the other hand, is linked to severe deterioration and poor performance. The procedure also offers a number of dangers by allowing lithium plating on the negative electrode. In light of these consequences, it is suggested that alternate charging techniques be utilized during high-rate charging to reduce deterioration and extend cell life.

Experiment results show that charging protocol optimization is another key area of fast-charging protocols. Charging protocol optimization is achieved through charging and discharging. [21] shows in his experiment that the charging process affects the SEI layer thickness and the cell capacity fade, whereby the ratio of the two rises as the cells are charged. On the contrary, the discharge has no impact on SEI and capacity fade [21]. When SEI increases and lithium plating occurs, capacity fade is experienced in the negative electrode of the cell. The effects of lithium plating cyclable lithium are considered to be minimal than SEI impact. Despite this, lithium plating remains a safety concern because it can cause short-circuiting in the cell. The charging process is only considered safe when the electrode potential is higher than the electrolyte potential. This scenario implies that the SEI potential is positive. Maintaining the SEI potential positive is a challenge in fast charging because fast rate charging increases the SOC, consequently decreasing SEI potential creating conditions for plating.

The study is significant in developing fast-charging protocols for Li-Ion batteries. Li-Ion batteries are promising in making electric vehicle technology a global reality by addressing the current charging limitations affecting its adoption. The study identifies various fast-charging protocols for these batteries and the level to which their effectiveness has been tested. Reviewing the protocols is vital as it provides insights to manufacturers on what to invest in and the constraints that need to be addressed. The study's significance is also because it addresses the effects of different charging protocols on degradation and how it can be minimized to lengthen the cell life. Literature used in the research further provides recommendations for minimizing and eliminating problems that cause cell degradation, such as heating and lithium plating. Additionally, recommendations for future studies are provided, which are critical for improving the charging protocol mechanisms to meet the commercialization needs for Li-Ion batteries used in electric vehicles. Despite these benefits, there are significant knowledge gaps identified in the constraints section that might limit the effectiveness of the review in future applications.

2.5 Conclusion

The Li-Ion battery is considered the most efficient among the battery technologies. They are promising in meeting the electric vehicles commercialization needs. The Li-Ion batteries have a high energy density, low and falling cost and longer life than the counterparts. The functionality of Li-Ion batteries depends on various aspects like its internal configuration and charging protocols. The charging protocols that apply initially do not meet the commercialization needs of these batteries and their use in the electric vehicles. Therefore, a lot of investment and investigation is going on in developing the effective fast-charging protocols. Although fast charging is highly recommended, the existing fast-charging protocols have their pros and cons. They could affect the cell functionality by causing thermal runaway. The key disadvantage of the majority of existing protocols is a fast battery degradation and diminished life. Additionally, current protocols are associated with extreme heating that leads to high temperatures during charging. These factors raise safety and economic concerns that need to be addressed before the fast-charging protocols are adopted for commercial usage. Moreover, the fast-charging protocols will be used on high-voltage battery packs, which requires additional safety measures. Certain safety measures, suggested to be used in the fast-charging arrangements, include reliable time-monitoring strategies and thicker insulations. Manufacturers can also improve safety by considering recommendations like evaluating the critical issues associated with fast charging, before the commercialization. This will help in identifying and eliminating the safety concerns.

The literature reviewed in this study identifies the factors that contribute to lithium plating as this is one of the key challenges in developing the fast-changing schemes. It is recommended that the used material must have a fast kinetics. The graphite remains the most appropriate material for developing the negative electrode. The lithium-titanate-oxide is also recommended for its high kinetics potential. However, further investigation is required to assess how the lithium-titanate-oxide's potential energy can be reduced. Li-Ion battery developers also need to invest in electrode designs that can accommodate fast diffusion. Current designs lack the capacity to facilitate the level of diffusion, which is required. Considering that the fast-charging protocols are new and still evolving, more studies are required to assess the impact of these protocols on various Li-Ion battery materials. It is evident that limited studies have addressed extreme fast-charging effects on materials. Most of these studies have focused on the graphite. Further studies should be conducted on the extreme fast-charging effects while considering the materials like silicon and lithium metal. The effects should also be assessed in electrolytes and separators to help in enhancing the cell efficiency by using materials that are least affected. Studies and improvements on methods that can be used to measure the cell parameters in real time are also needed. Also in the context of fast charging, the physical parameters of electrode play a significant role in the life and sensitivity of the battery. Therefore, it is recommended to investigate thoroughly the electrode design, which can accommodate the need for fast charging. Another key recommendation is to develop effective control and management algorithms. Cell arrangement will be different in fast-charging packs than it is in current Li-Ion batteries. Manufacturers have recommended that these batteries will have more cells in series than in parallel. This form of arrangement is beneficial as it allows better control, management and easy fault detection during charging. Additionally, future studies should be conducted to assess the cost-management strategies that can be utilized as fast-charging packs get commercialized.

In future studies the potential applications and protocols specific to the power system operation will also be investigated. Another future work is to conduct a critical analysis of the fast-charging protocols by following by international standards such as the Institute of Electrical and Electronics Engineers (IEEE), American National Standards Institute (ANSI) and British Standards (BS). Moreover, another prospect is the analysis of fast-charging circuits and applications in distribution systems.

Acknowledgements

The Effat University funds this work under grant number Decision No. UC#9/2June2021/7.2-21(3)5.

References

1. D. Deng, "Li-ion batteries: basics, progress, and challenges," Energy Sci. Eng., vol. 3, no. 5, pp. 385-418, 2015.
2. N. T. Al-Sharify, D. Raheem Rzaij, Z. Majid Nahi, and Z. T. Al-Sharify, "An experimental investigation to redesign A pacemaker training board for educational purposes," IOP Conf. Ser. Mater. Sci. Eng., vol. 870, p. 012020, 2020.
3. C. Gebhardt, "Multi-spacewalk series to replace Station batteries completed," Nasaspaceflight.com, 20-Jan-2020. [Online]. Available: https://www.nasaspaceflight.com/2020/01/us-segment-five-spacewalk-series-iss-batteries/. [Accessed: 2 Dec. 2020].
4. P. Zhang, J. Liang, and F. Zhang, "An overview of different approaches for battery lifetime prediction," IOP Conf. Ser. Mater. Sci. Eng., vol. 199, p. 012134, 2017.
5. "Battery Lifetime and Maintenance," Pveducation.org. [Online]. Available: https://www.pveducation.org/pvcdrom/battery-characteristics/battery-lifetime-and-maintenance. [Accessed: 2 Dec. 2020].
6. N. P. Ankit Gupta, "Lithium Ion Battery Market by Chemistry," Global Market Insights, May-2020. [Online]. Available: www.gminsights.com. [Accessed: 2 Dec. 2020].
7. Y. Mekonnen, H. Aburbu, and A. Sarwat, "Life cycle prediction of Sealed Lead Acid batteries based on a Weibull model," J. Energy Storage, vol. 18, pp. 467-475, 2018.
8. Tomaszewska, A. et al. (2019). Lithium-ion battery fast charging: A review. eTransportation. DOI: 10.1016/j.etran.2019.100011
9. Khan, H., Nizami, I. F., Qaisar, S. M., Waqar, A., & Krichen, M., "Analyzing Optimal Battery Sizing in Microgrids Based on the Feature Selection and Machine Learning Approaches", Preprint, 2022.

10. T. Chen *et al.*, "Applications of lithium-ion batteries in grid-scale energy storage systems," Trans. Tianjin Univ., vol. 26, no. 3, pp. 208-217, 2020
11. M. Ghiji *et al.*, "A review of lithium-ion battery fire suppression," Energies, vol. 13, no. 19, p. 5117, 2020.
12. Sheng, J., Tong, S., He, Z., & Yang, R. (2017). Recent developments of cellulose materials for lithium-ion battery separators. Cellulose, 24, 4103-4122.
13. Mian Qaisar, S., AbdelGawad, A. E. E., & Srinivasan, K., "Event-Driven Acquisition and Machine-Learning-Based Efficient Prediction of the Li-Ion Battery Capacity", SN Computer Science, 3(1), 1-6, 2022.
14. S. Bhattacharya, A. R. Riahi, and A. T. Alpas, "Thermal cycling induced capacity enhancement of graphite anodes in lithium-ion cells," Carbon N. Y., vol. 67, pp. 592-606, 2014.
15. Yang X-G, Zhang G, Ge S, & Wang C-Y. (2018). Fast charging of lithium-ion batteries at all temperatures. Proc. Natl. Acad. Sci. U.S.A, 115(28):7266e71. https://doi.org/10.1073/pnas.1807115115.
16. Gallagher KG, Trask SE, Bauer C, Woehrle T, Lux SF, Tschech M, L. P, Polzin BJ, Ha S, Long B, Wu Q, Lu W, Dees DW, & Jansen AN. (2016). Optimizing areal capacities through understanding the limitations of lithium-ion electrodes. J Electrochem 163(2):138e49. https://doi.org/10.1149/2.0321602jes
17. Waldmann T, Kasper M, & Wohlfahrt-Mehrens M. (2015). Optimization of charging strategy by prevention of lithium deposition on anodes in high-energy lithium-ion batteries electrochemical experiments. Electrochim Acta, 178:525e32. https://doi.org/10.1016/J.ELECTACTA.2015.08.056.
18. Yin, Y., Hu, Y., Choe, S., Cho, H., & Joe, T.W. (2019). New fast charging method of lithium-ion batteries based on a reduced order electrochemical model considering side reaction. Journal of Power Sources, 423, 367-379.
19. Qaisar, S. M., & AbdelGawad, A. E. E., "Prediction of the Li-Ion Battery Capacity by Using Event-Driven Acquisition and Machine Learning". In 7th International Conference on Event-Based Control, Communication, and Signal Processing (EBCCSP) (pp. 1-6). IEEE, 2021.
20. Xu, M. *et al.* (2018). Modeling the effect of two-stage fast charging protocol on thermal behavior and charging energy efficiency of lithium-ion batteries. Journal of Energy Storage, 20, 298-309.
21. Xu *et al.* (2019). Fast charging optimization for lithium-ion batteries based on dynamic programming algorithm and electrochemical-thermal-capacity fade coupled model. Journal of Power Sources, 438.
22. Keil, P., & Jossen, A. (2016). Charging protocols for lithium-ion batteries and their impact on cycle life—An experimental study with different 18650 high-power cells. Journal of Energy Storage, 6, 125-141.
23. Zhang S, Xu K, & Jow T. (2006). Study of the charging process of a LiCoO2-based Li-Ion battery. J Power Sources, 160(2):1349e54. https://doi.org/10.1016/J.JPOWSOUR.2006.02.087.

24. Qaisar, S. M., & Alguthami, M., "An Effective Li-Ion Battery State of Health Estimation Based on Event-Driven Processing". Green Energy: Solar Energy, Photovoltaics, and Smart Cities, 167-190, 2020.
25. Liu, K., Li, K., Peng, Q., & Zhang, C. (2019). A brief review on key technologies in the battery management system of electric vehicles. Frontiers of Mechanical Engineering, 14(1), 47-64.
26. Mian Qaisar, S., "Event-driven coulomb counting for effective online approximation of Li-ion battery state of charge". Energies, 13(21), 5600, 2020.
27. Fijita, Yoshikazu et al. "Development Of Battery Management System". Fujitsu Ten Technical Journal, vol. 42, no. 2016, 2016, pp. 68-80, https://www.denso-ten.com/business/technicaljournal/pdf/42-10.pdf. Accessed 20 Feb 2021.
28. Xing, Yinjiao et al. "Battery Management Systems In Electric And Hybrid Vehicles". Energies, vol 4, no. 11, 2011, pp. 1840-1857. MDPI AG, doi:10.3390/en4111840. Accessed 20 Feb. 2021.
29. Qaisar, S. M., "Li-Ion Battery SoH Estimation Based on the Event-Driven Sampling of Cell Voltage". In 2nd International Conference on Computer and Information Sciences (ICCIS) (pp. 1-4). IEEE, 2020.
30. Birkl, C. R., Roberts, M. R., McTurk, E., Bruce, P. G., & Howey, D. A. (2017). Degradation diagnostics for lithium ion cells. Journal of Power Sources, 341, 373-386.
31. Bach, T. C., Schuster, S. F., Fleder, E., Müller, J., Brand, M. J., Lorrmann, H., ... & Sextl, G. (2016). Nonlinear aging of cylindrical lithium-ion cells linked to heterogeneous compression. Journal of Energy Storage, 5, 212-223.
32. Yang, X. G., Leng, Y., Zhang, G., Ge, S., & Wang, C. Y. (2017). Modeling of lithium plating induced aging of lithium-ion batteries: Transition from linear to nonlinear aging. Journal of Power Sources, 360, 28-40.
33. Park, J., Appiah, W. A., Byun, S., Jin, D., Ryou, M. H., & Lee, Y. M. (2017). Semi-empirical long-term cycle life model coupled with an electrolyte depletion function for large-format graphite/LiFePO4 lithium-ion batteries. Journal of Power Sources, 365, 257-265.
34. Mian Qaisar, S., "A proficient Li-ion battery state of charge estimation based on event-driven processing". Journal of Electrical Engineering & Technology, 15(4), 1871-1877, 2020.
35. Han, X., Lu, L., Zheng, Y., Feng, X., Li, Z., Li, J., & Ouyang, M. (2019). A review on the key issues of the lithium ion battery degradation among the whole life cycle. ETransportation, 1, 100005.
36. B. Balagopal, C. S. Huang, and M.-Y. Chow, "Effect of calendar aging on li ion battery degradation and SOH," in IECON 2017 - 43rd Annual Conference of the IEEE Industrial Electronics Society, 2017.
37. S. Ma et al., "Temperature effect and thermal impact in lithium-ion batteries: A review," Prog. Nat. Sci., vol. 28, no. 6, pp. 653-666, 2018.

38. Abdi, H., Mohammadi-ivatloo, B., Javadi, S., Khodaei, A. R., & Dehnavi, E. (2017). Energy storage systems. Distributed Generation Systems, 333-368.
39. Vetter, J., Novák, P., Wagner, M. R., Veit, C., Möller, K. C., Besenhard, J. O., ... & Hammouche, A. (2005). Ageing mechanisms in lithium-ion batteries. Journal of Power Sources, 147(1-2), 269-281.
40. Zeng, X., Xu, G. L., Li, Y., Luo, X., Maglia, F., Bauer, C., ... & Chen, Z. (2016). Kinetic study of parasitic reactions in lithium-ion batteries: a case study on LiNi0. 6Mn0. 2Co0. 2O2. ACS Applied Materials & Interfaces, 8(5), 3446-3451.
41. Qaisar, S. M., "Event-Driven Approach for an Efficient Coulomb Counting Based Li-Ion Battery State of Charge Estimation". Procedia Computer Science, 168, 202-209, 2020.
42. Lu, B., Ning, C., Shi, D., Zhao, Y., & Zhang, J. (2020). Review on electrode-level fracture in lithium-ion batteries. Chinese Physics B, 29(2), 026201.
43. Tippmann, S., Walper, D., Balboa, L., Spier, B., & Bessler, W. G. (2014). Low-temperature charging of lithium-ion cells part I: Electrochemical modeling and experimental investigation of degradation behavior. Journal of Power Sources, 252, 305-316.
44. Waldmann, T., Wilka, M., Kasper, M., Fleischhammer, M., & Wohlfahrt-Mehrens, M. (2014). Temperature dependent ageing mechanisms in Lithium-ion batteries–A Post-Mortem study. Journal of Power Sources, 262, 129-135.
45. Pinson, M. B., & Bazant, M. Z. (2012). Theory of SEI formation in rechargeable batteries: capacity fade, accelerated aging and lifetime prediction. Journal of the Electrochemical Society, 160(2), A243.
46. Prada, E., Di Domenico, D., Creff, Y., Bernard, J., Sauvant-Moynot, V., & Huet, F. (2013). A simplified electrochemical and thermal aging model of LiFePO4-graphite Li-Ion batteries: power and capacity fade simulations. Journal of the Electrochemical Society, 160(4), A616.
47. Mei, W., Chen, H., Sun, J., & Wang, Q. (2019). The effect of electrode design parameters on battery performance and optimization of electrode thickness based on the electrochemical–thermal coupling model. Sustainable Energy & Fuels, 3(1), 148–165. https://doi.org/10.1039/C8SE00503F
48. Zhao, R., Liu, J., & Gu, J. (2015). The effects of electrode thickness on the electrochemical and thermal characteristics of lithium ion battery. Applied Energy, 139, 220-229. https://doi.org/10.1016/j.apenergy.2014.11.051
49. Pender, J. P., Jha, G., Youn, D. H., Ziegler, J. M., Andoni, I., Choi, E. J., ... & Mullins, C. B. (2020). Electrode degradation in lithium-ion batteries. ACS Nano, 14(2), 1243-1295.
50. Liu, P., Wang, J., Hicks-Garner, J., Sherman, E., Soukiazian, S., Verbrugge, M., ... & Finamore, P. (2010). Aging mechanisms of LiFePO4 batteries deduced by electrochemical and structural analyses. Journal of the Electrochemical Society, 157(4), A499.

51. Sarre, G., Blanchard, P., & Broussely, M. (2004). Aging of lithium-ion batteries. Journal of Power Sources, 127(1-2), 65-71.
52. Amine, K. (2002). Factors responsible for impedance rise in high power lithium ion batteries. In Fuel and Energy Abstracts (Vol. 4, No. 43, pp. 261-262).
53. Groot, J. (2012). State-of-health estimation of Li-Ion batteries: Cycle life test methods.
54. Plett, G. (2015). Battery Management Systems, Volume 1. Artech House.
55. Li, J., Zhou, S., & Han, Y. (2016). Advances in Battery Manufacturing, Service, and Management Systems. John Wiley & Sons.
56. AL-Refai, A., Rawashdeh, O., & Abousleiman, R. (2016). An Experimental Survey of Li-Ion Battery Charging Methods. International Journal of Alternative Powertrains, 5(1), 23-29. Retrieved 31 October 2020, from https://www.jstor.org/stable/26169102.
57. Weicker, P. (2014). A Systems Approach to Lithium-ion Battery Management. Artech House.
58. Xiong, R. (2019). Battery Management Algorithm for Electric Vehicles. Springer Nature.
59. Tomaszewska, A. et al. (2019). Lithium-ion battery fast charging: A review. eTransportation. DOI: 10.1016/j.etran.2019.100011
60. Zhang, G., Ge, S., Xu, T., Yang, X. G., Tian, H., & Wang, C. Y. (2016). Rapid self-heating and internal temperature sensing of lithium-ion batteries at low temperatures. Electrochimica Acta, 218, 149-155.
61. Ahmed, S. et al. (2017). Enabling fast charging - A battery technology gap assessment. Journal of Power Sources, 367, 250-262.
62. Computer Aided Engineering for Batteries. (2002). Community Battery Simulation Framework. http://batterysim.org.
63. U.S. Department of Energy. (2017). Computer Aided Engineering for Batteries, https://energy.gov/eere/vehicles/downloads/computer-aided-engineering-electric-drive-vehicle-batteries-caebat
64. Zhang S, Xu K, & Jow T. (2006). Study of the charging process of a LiCoO2-based Li-Ion battery. J Power Sources, 160(2):1349e54. https://doi.org/10.1016/J.JPOWSOUR.2006.02.087.
65. Dahn H. M., Smith A. J., Burns J. C., Stevens D. A. & Dahn J. R. (2012). J. Electrochem. Soc. 159, A1405
66. Attia, P. M., Grover, A., Jin, N., Severson, K. A., Markov, T. M., Liao, Y. H., ... & Chueh, W. C. (2020). Closed-loop optimization of fast-charging protocols for batteries with machine learning. Nature, 578(7795), 397-402.
67. Williard, N., He, W., Osterman, M., & Pecht, M. (2012, August). Reliability and failure analysis of Lithium Ion batteries for electronic systems. In 2012 13th International Conference on Electronic Packaging Technology & High Density Packaging (pp. 1051-1055). IEEE.
68. Cannarella, J. & Arnold, C.B. (2015). Electrochem. Soc. 162. A1365eA1373.

69. Burns, J.C., Stevens, D.A., & Dahn, J.R. (2015). J. Electrochem. Soc. 162, A959-A964.
70. Computer Aided Engineering for Batteries. (2002). Community Battery Simulation Framework. http://batterysim.org.

3

Development of Sustainable High-Performance Supercapacitor Electrodes from Biochar-Based Material

Kriti Shrivastava[1*] and Ankur Jain[1,2†]

[1]School of Applied Sciences, Suresh Gyan Vihar University, Jagatpura, Jaipur, India
[2]Centre for Renewable Energy & Storage, Suresh Gyan Vihar University, Jagatpura, Jaipur, India

Abstract

The future energy industry will be more electric, efficient, networked, and environment friendly as compared to the present one. Clean energy technology is rapidly becoming cost-effective, broadly usable, and a substantial new source of investment, job creation, and international collaboration, due to rapid technological breakthroughs and global legislative reforms. The increasing expansion of renewable generation resources necessitates the presence of energy storage systems which will ensure the energy security, and electrification of remote and rural areas and will also cut down the carbon emission. They can assure the reliability of electric power grids along with the continuous supply of electric power. In response to the rising need for energy-storage devices, supercapacitors, which are electrochemical capacitors, are being explored intensively. They have several advantages over standard secondary storage batteries, including faster charge propagation, which is intrinsically simpler and reversible, longer cycle life, and higher storage efficiency. In modern energy storage and conversion, carbon materials have the potential to be the most adaptive basic materials, and biochar, or bio carbon, can be easily chosen for commercial Electric Double Layer Capacitors (EDLCs) because of its larger surface area, high chemical stability, comparatively low cost, appropriate electrical conductivity, and availability. Plants residues, agricultural by-products, and waste biomass can be effectively utilized for biomass production

*Corresponding author: kriti.shrivastava@mygyanvihar.com; https://orcid.org/0000-0001-9690-4124
†Corresponding author: ankur.j.ankur@gmail.com; https://orcid.org/0000-0002-9632-2682

Sandeep Dhundhara, Yajvender Pal Verma, and Ashwani Kumar (eds.) *Energy Storage Technologies in Grid Modernization*, (71–104) © 2023 Scrivener Publishing LLC

as a carbon-negative process. This provides an effective and sustainable method for the manufacture of supercapacitor electrodes with major benefits coming in terms of performance with economic and environmental aspects.

Keywords: Biochar, supercapacitor, electric double layer capacitor, electrode, sustainable electrode material

3.1 Introduction

The economic recovery from Covid-induced recession is taking place rapidly and it is putting strain on all aspects of today's energy system, which has resulted in a significant hike in the cost of traditional energy resources like natural gas, coal, and oil. Despite all the advances made in renewable energy and electric transportation, coal, and oil consumption increased dramatically in 2021, which resulted in a sharp increase in CO_2 emission. The energy requirement of the future will be more electric, efficient, networked, and environmentally benign than the present one. With rapid technological advancements and worldwide policy changes, sustainable energy systems are gradually becoming cost-effective and widely applicable. Clean energy technology is also attracting significant new investments and creating job opportunities by enhancing the international collaborations between government and non-government organizations [1].

The unprecedented changes in the electric power systems are reflected by increasing attention over the development of smart grids and grid modernization [2]. To address the energy and environmental crises caused by using fossil fuels, a sustainable energy system must be developed and implemented on a broad scale, which can only be accomplished through the development and implementation of renewable energy. The promise of energy storage application is steadily appearing as the smart grid develops, aided by investment and government legislation. These applications span the whole power system spectrum from power generation to power transmission, its distribution, and consumption. They can address increasing the utilization of renewable energy sources at large scale, decreasing the overall construction cost of generation, and improving the efficiency of the power grid system, power quality, and power supply. Such applications also perform grid optimization through planning, management, and operational control [3].

Conventional utilization of fossil fuel resources has raised environmental concerns over a very long time along with the issues of limited availability and energy security. Being inexhaustible and environment-friendly, renewable energy generation has gained worldwide attention. Solar and wind energy are the two widely available infinite sources of energy, but they are not constant and reliable. In such conditions, thermal power plants are required to be used to balance the fluctuation of these renewable sources and therefore electric grid operators face considerable hurdles in such applications [4]. Some other important applications of energy storage are the integration of grid generating renewable energy, distribution, distributed generation, microgrid, and its auxiliary services [3].

As compared to the other existing energy storage technologies, supercapacitors are associated with advantages like quick charge-discharge, good cycle stability, and high-power density. Carbon is the most adaptable platform material which can be suitably applied in modern sustainable energy storage systems. They can be used in the form of activated carbon, and carbon nanostructures like fibers, tubes, and graphene. But their preparation involves energy-intensive complex synthetic procedures using coal or petrochemical compounds as the precursor. Several studies had been conducted to develop sustainable eco-friendly methods for efficient production of carbon-based materials using renewable resources [5].

Biochar or the bio-carbon is characterized by the presence of several surface functional groups, and easily adjustable porosity which makes it a suitable choice as a sustainable material having considerable application promise in energy storage and conversion during various recent studies. When compared to the other carbon material, Biochar is found to be a more efficient and active low-cost material [6, 7].

This chapter covers detailed information on the suitability of biochar-derived carbon nanomaterials for preparing electrodes of low-cost and effective supercapacitors for sustainable energy storage technologies. The major advantage is the control over their electrochemical activity by suitable modifications in the textural properties and surface functional groups, variation in the feedstock sources, and manufacturing procedures. This provides an effective method to the manufacture of supercapacitor electrodes which are economic, eco-friendly, and able to give high performance.

3.2 Role of Energy Storage Systems in Grid Modernization

Energy storage systems involve three major steps in energy storage [8]:

1. The charge is the process of receiving electrical energy from the source.

2. Storage is the process of conversion of electrical energy into some other form and its storage.
3. Discharge is the process of receiving the stored electrical energy from the system.

Furthermore, all the storage systems consist of two parts:

(i) Central storage which is the repository of energy storage performs the power conversion.
(ii) Being present between the central storage and the power system, the interface controls the bidirectional transfer and the level of charge or discharge of the stored energy by employing measuring devices like sensors. Because energy storage is merely a depository of energy rather than an ideal energy source, it is accompanied with energy losses at every stage of the storage process [9].

Energy storage is the most significant aspect in improving the electrical energy grid's flexibility, economy, and security. As a result, energy storage is projected to support distributed power and microgrids, as well as encourage open sharing and flexible trading of energy production and consumption and achieve multi-functional coordination [6].

Electrical energy storage (EES) is an important part of the smart and modern future grid, which should be able to hold a huge amount of renewable energy. Generally, renewable energy resources are variable, intermittent, and located far away from the load centers. To tackle all these issues, an effective electrical energy storage system is required. It will also be beneficial for fueling hybrid and electric vehicles (i.e., electricity), even though the cost of establishing EES is a major problem [4].

In principle, an energy storage systems (ESS) should focus on its energy storage capacity during the time when it is easily available and its release during the period of its shortage. It should provide stability to the power output by combining the energy storage technology with renewable energy resources to enhance its reliability. It should also focus on increasing the resilience of the power system during natural disasters, weather variations, and the sudden onset of power shortage [10, 11].

In commercial applications, EES technology should be able to maintain a continuous and reliable power supply [12]. Numerous energy storage technologies are available to cater to the need of a modern electrical grid system (Figure 3.1). All these technologies can be classified into two broad

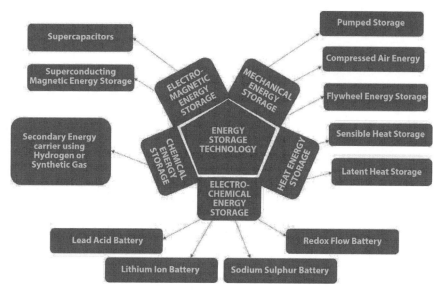

Figure 3.1 Various energy storage technologies.

categories on the basis of mode of electrical energy storage. Supercapacitors store electrical energy directly in the form of electrical charge and they are highly efficient due to their low energy density and shorter discharge time. Currently, they are being utilized for frequency regulation in power management.

Alternatively, electrical energy can be converted into any of the other forms of energy like kinetic energy, potential energy, or chemical energy and afterwards it can be stored (Table 3.1). Mechanical energy storage systems like flywheels (FWs) store electrical energy by high-speed rotors spinning with their speed directly proportional to stored energy [13]. When the speed of rotors is slowed down, the mechanical energy can be converted again into electrical energy offering high power but low energy. They are also useful for power management. For bulk energy storage, electrical energy is stored in the form of potential energy by the pumped hydro and compressed air energy storage (CAES).

Most of the existing technologies are associated with the disadvantage of high cost due to the expenditure pertaining to operation, maintenance, replacement, and additional carrying costs. It can also be attributed to their unsatisfactory performance, high cost of raw materials and fabrication and the scale of production. To overcome this problem, technological

Table 3.1 Techno-economic parameters of various energy storage devices [9]. *Reprinted with permission of Elsevier.*

Technologies		Capacity (MWh)	Power (MWh)	Response time	Discharge time	Maturity	Lifetime (Years)	Efficiency (%)	Advantages	Disadvantages
Electrochemical	Lead-Acid	0.25-50	≤ 100	millisecond	≤ 4 h	Demo-Commercial	≤ 20	≤ 85	Inexpensive, High recyclable, Reality available	Very heavy, Limited usable energy; Poor energy density
	Lithium Ion	0.25-25	≤ 100		≤ 1 h	Demo	≤ 15	≤ 90	High Capacity; Great Stability in calendar and cycle life	
	NaS	≤ 300	≤ 50		≤ 6 h	Commercial	≤ 15	≤ 80	High storage capacity, Inexpensive	Working only when the sodium and sulfur are liquids 290-390°C
	Vanadium Redox	≤ 250	≤ 50	≤ 10 min	≤ 8 h	Demo	≤ 10	≤ 80	Possible to use for many different renewable energy sources	
Mechanical	FES	≤ 10	≤ 20	≤ 10 ms	≤ 1 h	Demo-mature	≤ 20	≤ 85	High power density, non-polluting, high efficiency	not safe enough, Noisy; High-speed operation led to vibration
	PHS (Small)	≤ 5000	≤ 500	sec-min	6-24 hrs	mature	≤ 70	≤ 85	Remote operation is possible, low manpower factor, relatively low maintenance	Silt build-up, Impedance to the movement of environmental issues
	PHS (Large)	≤ 14000	≤ 1400	sec-min						

(Continued)

Table 3.1 Techno-economic parameters of various energy storage devices [9]. Reprinted with permission of Elsevier. (Continued)

Technologies		Capacity (MWh)	Power (MWh)	Response time	Discharge time	Maturity	Lifetime (Years)	Efficiency (%)	Advantages	Disadvantages
	CAES (underground, small)	≤ 1100	≤ 135	≤ 15 min	≤ 8 h	Demo-Commercial	≤ 40	≤ 85	High power capacity, low losses (can be storage energy for more than a year) Fast Start-up	It is not possible to install everywhere, and the location is dependent on a geological structure
	CAES (underground, large)	≤ 2700	≤ 135	≤ 15 min	≤ 20 h	Demo				
	CAES (above ground)	≤ 250	≤ 50	≤ 15 min	≤ 5 h					
Electrical	DLC	0.1–0.5	≤ 1	≤ 10 ms	≤ 1 min	Commercial	≤ 40	≤ 95	High power density, low resistance, high efficiency	Low energy density, low voltage per cell, incomplete capacity utilization
	SMES	1.0–3.0	≤ 10	≤ 10 ms	≤ 1 min	Commercial	≤ 40	≤ 95	High power, high efficiency, environmentally safe	For sizing of high energy storage need to long loop, cooling system is needed, expensive
Thermal	Thermal	≤ 350	≤ 50	≤ 10 min	N/A	Mature	≤ 40	≤ 90	Non-polluting unlimited energy source	Expensive; depends on a geological structure

advancements are required to improve the reliability, cycle life, and efficiency of conversion by using less expensive materials [4].

Carbon-free source with operational adaptability can be utilized to facilitate the integration of multiple renewable energy sources and for the improvement of the usage of generation assets. Therefore, energy storage systems (EES) can help in attaining carbon neutrality in the power sector. EES could help generators to use less fossil fuel, resulting in lower standard discharge and GHG emissions [14]. This includes emissions from the entire energy production chain right from generation to transmission and distribution to the final user. Commercially modern energy storage can also help in the transition from centralized to more advantageous distributed generation system to improve energy access, availability in rural and remote locations along with the quality, dependability, and performance [13].

3.3 Overview of Current Developments of Supercapacitor-Based Electrical Energy Storage Technologies

The growing demand of efficient energy storage devices has caused extensive research on supercapacitors due to their advantages over traditional batteries, e.g., rapid charge propagation, extended cycle life, and higher storage efficiency. Due to these reasons, they can be easily applied in energy storage for electric vehicles, UPS (uninterrupted power supply), digital telecommunication and pulse lasers, etc. Although supercapacitors are known for their high power density, quick reaction, cheap maintenance, and wide range of operating temperature, they are usually employed in combination with other energy storage technologies [3, 11].

Various materials have been explored as potential electrode materials for supercapacitors and based on material selection supercapacitors can be classified into two parts: pseudocapacitors and electric double-layer supercapacitors (Figure 3.2). Pseudocapacitors generally contain materials like hydroxides, oxides, and polymers and exhibit improved specific capacitance and high energy density. However, mostly they depend on faradic redox processes and the active materials behave as insulators to slow down fast electron transport, which is necessary for higher rates. Therefore, these pseudocapacitors frequently compromise rate capability and reversibility [15].

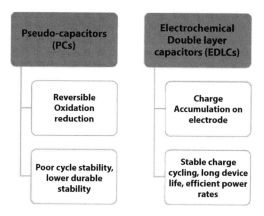

Figure 3.2 Types of supercapacitors.

On the other hand, electrochemical double-layer capacitors (EDLC) or ultracapacitors can efficiently replace batteries in energy storage systems and harvesting applications, which requires a high-rate power delivery and uptake. Small-scale supercapacitors can alone be utilized as power sources for microelectronic devices or they can be integrated with batteries and other energy harvesters for wider industrial applications [16].

The electrochemical double-layer capacitors store electrical energy in an electrolytic double layer produced due to the separation of charged species, and this charge storage mechanism is fundamentally simpler and more reversible than secondary storage batteries. EDLCs' electrode/electrolyte interfaces are perfectly polarizable and are not affected by faradaic reactions over the entire potential range of operation. But major problems associated with them is their restricted cycle life due to the device packaging rather than by device component deterioration [17].

Large-scale commercial utilization of the carbon materials in supercapacitors is due to their low cost, easy processability, and versatile existing forms (Figure 3.3). They follow charging–discharging through the electric double-layer mechanism. The carbon-based supercapacitor electrodes come in diverse shapes and sizes due to the numerous types of available carbon compounds. The thin-film electrodes are the most common and can be easily prepared by combining carbon compounds with a binder to form a slurry, and then printed or rolled into film electrodes [18]. Because of their chemical inertness and ease of processing, carbonaceous materials have piqued researchers' interest in using them as supercapacitor electrodes in recent years. The well-developed surface area, size and shape distribution of pores, average pore size, wettability, conductivity, and presence

Figure 3.3 Advantages of carbon material for electrochemical energy storage.

of electroactive species are all important aspects while selecting suitable carbon materials for capacitor electrodes. Due to its amphoteric nature, carbon displays high electrochemical characteristics both as donor and acceptor, making it an appealing material for supercapacitors. Availability of enormous surface area, low density, and high electrical conductivity combined with exceptional chemical stability contribute to their excellent capacity to accumulate charges [19].

3.4 Potential of Biochar as High-Performance Sustainable Material

Carbon materials have the potential to be the most adaptable base materials in modern energy storage and conversion. When such compounds are derived from coal and petrochemicals, it is typically energy-intensive to generate or involve difficult conditions for synthesis. The development of efficient ways for producing carbon materials from its renewable resources with excellent performance characteristics and minimal environmental impact is a top priority. Biochar is a type of bio-carbon with numerous surface functional groups and easily controllable porosity. Therefore it could prove to be an excellent candidate as a long-term carbon material in this case [7].

Recent studies show great applications of biochar-based materials in energy storage and conversion due to adaptable surface properties. Carbon aerogels have enormous (specific) capacities and capacitance densities due to their unique nanostructure. Because of the carbon electrode's continuous structure, this stored energy may be released quickly and efficiently [17].

Generally activated carbon and carbon nanostructures of commercial importance are prepared from coal and petrochemicals via energy-intensive harsh synthetic processes. Different activation processes like steam activation, CO_2 activation, and chemical activation by acid, base or salt are used to prepare activated carbon from conventional fossil fuel resources [20]. On the other hand, specific techniques like chemical vapor deposition and electric arc discharge are required for the synthesis of carbon nanomaterials, which require high temperature and complex operational procedures [21–23]. As a result, traditional methods are ineffective for large-scale manufacture and commercial use of functional carbon-based materials. New effective procedures must be developed to generate high-performance carbon materials with little environmental impact. Biomass is a viable raw material for the production of different biochar-based compounds in the context of sustainable carbon materials (biochar) production due to its renewable nature and natural availability [7].

Biochar is basically a solid residue that is formed by the thermal breakdown of biomass at moderate temperatures (350-700°C) either in the absence or limited supply of oxygen [24]. It can be considered as a specific type of biocarbon obtained from plant biomass through the methods of pyrolysis, carbonization, hydrothermal treatment, and gasification. Based on synthesis methods, biochar can be distinguished from other carbon compounds. Generally, carbon black and activated carbon are synthesized from fossil fuels whereas charcoal can be obtained from any biological resources like plants, animals, or microorganisms (Figure 3.4). Biochar and activated carbon are characterized by a porous amorphous carbon matrix but the surface of biochar is frequently packed with functional groups like C–O, C=O, -COOH, and -OH, etc. [25]. Therefore, as compared to the activated carbon, biochar shows high adaptability and serves as a framework for the creation of diverse functionalized carbon compounds [26].

Large-scale biochar production should be considered as a sustainable process as compared to the carbon black and activated carbon prepared from fossil fuels because it reduces anthropogenic CO_2 emissions. Atmospheric CO_2 is fixed into biomass by plants and becomes a part of the carbon matrix of biochar. This is a cost-effective way to reduce CO_2 in the carbon cycle, hence slowing down global warming. By storing carbon

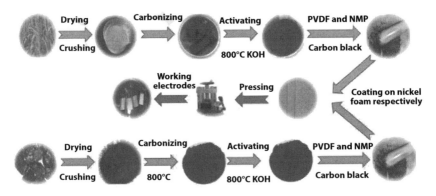

Figure 3.4 Preparation of modified activated carbon electrode [27]. Reprinted under creative commons license

in biochar, billions of tons of CO_2 can be removed from the carbon cycle [28, 29].

Energy storage and conversion technology is based on the combination of physical and chemical interactions occurring at the interface. Therefore biochar-based functional materials are particularly advantageous due to their easily tuned porosity and other surface properties. Functional groups on biochar surface enable the control over the thermodynamic features of surface interactions at the interface, while surface porosity can be changed according to the desired rate and kinetics of physical interactions and chemical reactions. Meanwhile, biochar's easily adjustable surface characteristics make it a versatile platform for the development of a variety of nano-structured biochar composites which can be used for energy storage and conversion.

Biochar-based nanocomposites have been widely used as supercapacitor electrode materials. Raw biochar materials are useful because, in comparison to typical carbon sources, oxygenated functional groups are often abundant, e.g., -OH, >C=O, and -COOH. By adjusting the pyrolysis parameters, such as temperature and heating rate, the oxygen concentration can be fine-tuned [30]. Furthermore, appropriate selection of suitable biomass precursor, adjusting pyrolysis temperature and catalyst, the quantitative energy storage capacity of the supercapacitor can be increased. By enhancing the pore access and surface utilization, the wettability of the carbon electrode can be improved by some electrochemically inert oxygenated functional groups and, as a result, the specific capacitance can also be increased. In addition to this, when highly oxygenated functional groups are present on the surface, they can inhibit further oxidation of the carbon matrix over a varied potential range which improves the electrode materials' cycle stability [31].

3.5 Overview of Recent Developments in Biochar-Based EDLC Supercapacitor

In the electrochemical double-layer capacitors (EDLCs), the amount of energy stored can be controlled through the charge separation at electrode/electrolyte interfaces (Figure 3.5). The charge separation distance is reduced to the dimensions equivalent to those of the ions within the electrolyte (few nanometers). This modification together with the availability of enormous surface area at the porous electrode surface supports the storage of much more energy than conventional capacitors. In comparison to batteries, the electrostatic mechanism of energy storage results in higher power densities, excellent reversibility, and longer cycle life. High surface area, easy availability, low cost, appropriate electrical conductivity, and chemical stability make biochar-based functional materials viable for commercial EDLCs.

In recent years, due to chemical inertness and ease of processing, carbonaceous materials have piqued researchers' interest in their application as supercapacitor electrodes. For the preparation of supercapacitor electrodes, several biochar precursors have been utilized. Broadly they can be classified into three categories:

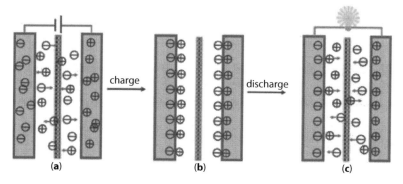

Figure 3.5 Mechanism of EDLC supercapacitors (a) the charging process, (b) EDLC after charging, and (c) the discharging process [32]. Reprinted under creative commons license.

1. Biochar-based supercapacitors derived from wood and other plant residues.
2. Biochar-based supercapacitors derived from waste biomass (industry, agriculture, solid waste).
3. Biochar-based supercapacitors derived from other methods.

3.5.1 Wood & Plant Residues as Biochar Precursor for Supercapacitor Applications

Manufacturing of carbon materials from biomass is a cheap sustainable technology due to the abundancy of biomass. Biochar is a valuable product obtained from biomass through series of chemical reactions and this process of converting biomass to biochar is considered carbon negative. Various morphologies of biochar like fiber, sheet and honeycomb can be obtained from biomass. Plants are a rich source of lignin and cellulosic material which is proved to be a superior precursor of biomass (Table 3.2) with several advantages like high surface area, extensive pore structure, variety of surface functional groups and able to attain superior conductivity after appropriate treatment [33].

Woody biochar is a promising eco-friendly low-cost material due to its excellent performance as electrode with low environmental impacts for supercapacitor applications. Jiang *et al.* [5] carbonized red cedar wood via one pot pyrolysis and prepared woody biochar monolith having ultra-high carbon content and extensively ordered macropores. Structural studies by EDX and SEM revealed very high carbon content of approximately 98% along with a highly ordered microporous texture. When chemical activation of the biochar was performed in dilute nitric acid at room temperature, a significant 115 $F.g^{-1}$ of capacitance was recorded which was seven times higher and this was attributed to the consideration due to the increase in the density of surface oxygen groups after activation.

Similarly, biochar was prepared by maple wood pyrolysis and three kinds of supercapacitor electrodes were prepared: mini-chunk, thin-film, and large-disk-chunk. The mini-chunk electrodes and thin film electrodes both showed outstanding performance characteristics and electrochemical behavior. The high specific capacitance (32 $F.g^{-1}$) and high stability of mini-chunk supercapacitor indicate their suitability to be used as small-scale power source in different electronic devices. Furthermore, the mini-chunk electrode allows for a quick and easy evaluation of biochar materials as possible high-performance, low-cost, and environmentally friendly supercapacitor electrodes without using a binder or costly production techniques [16].

Table 3.2 Wood & plant residues as biochar precursor for supercapacitor applications [7, 12, 40].

Biomass	Activation process	Surface area SBET ($m^2 \cdot g^{-1}$)	Specific capacitance ($F \cdot g^{-1}$)	Current density	Electrolyte used
Coconut kernel	KOH	1200	173	10 A g^{-1}	H_2SO_4
Microalgae	Hydrothermal KOH	1800-2000	200	0.25 A g^{-1}	LiCl
Willow catkins	Carbonization KOH	1775.7	292	0.1 A g^{-1}	KOH
Tea Leaves	Activation analysis	911.92	167	1 A g^{-1}	KOH
Ginko Leaves	Pyrolysis KOH	1775	178	1 A g^{-1}	KOH
Broad Bean	KOH	655.4	202	0.1 A g^{-1}	KOH
Wild rice stem	Hydrothermal KOH	1228.8	301	1 A g^{-1}	
Bamboo	Hydrothermal KOH	1472	301	0.1 A g^{-1}	KOH
Coffee Endocarp	Pyrolysis KOH, CO_2, HNO_3	89-1050	176	10 mA	H_2SO_4
Corn Stalk	KOH	2139	317	1 mVs-1	KOH
Moringa leaves	Carbonization	1327	323	1 A g^{-1}	KOH
Watermelon	KOH		358	50 A g^{-1}	KOH
Soybean root	KOH	2143	276	0.5 A g^{-1}	KOH
Aloe vera	KOH	1890	410	0.5 A g^{-1}	H_2SO_4
Hyacinth	KOH	2276	344.9	0.5 A g^{-1}	H_2SO_4
Tobacco	Hydrothermal KOH	2115	286.6	0.5 A g^{-1}	Aqueous
Elm samara	Pyrolysis KOH	1947	470	1 A g^{-1}	KOH (6 M)

(Continued)

Table 3.2 Wood & plant residues as biochar precursor for supercapacitor applications [7, 12, 40]. (*Continued*)

Biomass	Activation process	Surface area SBET (m².g⁻¹)	Specific capacitance (F.g⁻¹)	Current density	Electrolyte used
Bamboo	Hydrothermal KOH	1472	301	0.1 A g⁻¹	KOH (6 M)
Hemp stems	Pyrolysis Steam MnO_2	438	340	1 A g⁻¹	Na_2SO_4 (1 M)
Watermelon	Hydrothermal $FeCl_3$ $FeSO_4$		358	0.5 A g⁻¹	KOH (6 M)
Soybean roots	Pyrolysis KOH	2143	276	0.5 A g⁻¹	KOH (6 M)
Aloe vera	Pyrolysis KOH	1890	410	0.5 A g⁻¹	H_2SO_4 (1 M)
Typha angustifolia	Pyrolysis KOH	3062	257	0.5 A g⁻¹	K_2SO_4 (0.5 M)
Seaweeds	Pyrolysis	3487	203	0.05 A g⁻¹	H_2SO_4 (1 M)
Neem leaves	Pyrolysis	1230	400	0.5 A g⁻¹	H_2SO_4 (1 M)
Sisal	Pyrolysis KOH	2289	415	0.5 A g⁻¹	KOH
Rose flower	Pyrolysis KOH/ KNO_3	1980	350	1 A g⁻¹	KOH/ KNO_3
poplar anthers	Pyrolysis KOH	3639	361.5	0.5 A g⁻¹	KOH
shaddock endothelium	Pyrolysis KOH	1265	550	0.2 A g⁻¹	KOH
cornstalk	Pyrolysis $K_2C_2O_4 \cdot H_2O$	2054	461	1 A g⁻¹	$K_2C_2O_4 \cdot H_2O$
cellulose	Pyrolysis NaOH	1588	288	0.5 A g⁻¹	NaOH

Pinecones biochar-based activated carbon with high specific surface area (up to 2450 $m^2.g^{-1}$) and pore structure developed especially for the adsorption of redox-active polyoxometalate (POM) clusters. This material was used for impregnation of the $PMo_{12}O_{40}^{3-}$ (PMo_{12}) and this hybrid material thus prepared possesses high redox activity leading to high areal capacitance of 1.19 $F.cm^{-2}$ greater than the unmodified carbon material. This could be a potential way for designing high-performance hybrid energy storage electrodes at a low cost [34].

Eco-friendly wood-derived biochar (WDB) was prepared by Wan et al. [35]. They pyrolyzed wooden waste of agriculture and industry and then MnO_2/WDB composite was prepared by in situ redox reaction with $KMnO_4$. The core-shell structure thus prepared could be used as a free-standing, binder-free electrode with a specific capacitance of 101 $F.g^{-1}$, coulombic efficiency of 98%–100%, and enhanced cyclic stability along with a capacitance retention of 85.0% after 10,000 cycles. By blending it with other electrochemical active compounds, it can be employed as a unique, safe substrate material to produce high-performance energy storage devices.

High-temperature biochar supercapacitor was prepared with binder free biochar monolith and 1-butyl-3-methylimidazolium tetrafluoroborate-based liquid ionic electrolyte. It was discovered that raising the temperature up to 140°C could enhance the specific mass capacity and charge-discharge rate by ten times. Within a voltage range of 6V, cycle stability was obtained up to 1,000 cycles. Such supercapacitor electrodes have potential to be used in higher energy and power density energy storage devices [36].

A high-performance supercapacitor electrode was fabricated from the activated biochar of infested ash tree residue. Carbonization was done at 700°C to obtain the biochar which was chemically activated in an Ag_2SO_4/HNO_3 solution. This activated Ag/BC composite product was electrochemically compared with activated biochar (a-BC) chemically activated with HNO_3. A 31.4% higher specific capacitance was obtained for Ag/BC composite electrode. Also, high cycling stability was demonstrated with specific capacitance retention of 98.6% after 2,000 cycles [37].

Y. Li et al. [38] utilized pomelo peel for the synthesis of super-hydrophilic biochar materials with microporous structure for supercapacitor application (Figure 3.6). Chemical activation was done by zinc nitrate (porogen) and urea (nitrogen source) which yielded biochar material with extensive microporous structure and a very large nitrogen and oxygen doping content (>20%). This method produced super-hydrophilicity to this functional material and enhanced the availability of the micropores for charge storage. Under moderate conditions of calcination, the

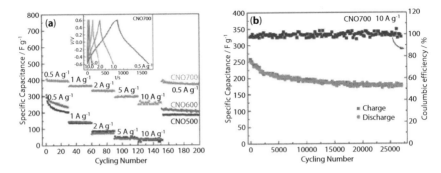

Figure 3.6 Galvanostatic charge–discharge performances of (a) all samples at different rates, and (b) CNO700 at 10 A g⁻¹ [38]. Reprinted with permission of Elsevier.

biochar material can be imparted high conductivity without decreasing the super-hydrophilicity. The material (CNO700) exhibits high and stable specific capacitance of 391.0 F. g^{-1} at 0.5 A. g^{-1} and cycling stability more than 25,000 times at 10.0 A. g^{-1} displaying good capacitance retention due to the large effective pore volume (0.69 cm^{3}g^{-1}) synergetic effect, considerable doping of heteroatoms with proper elemental ratio and high conductivity value of 0.713 S.m^{-1}.

Biochar can also serve as eco-friendly low-cost precursor material for the large-scale manufacture of carbon nano structures which can be used for future supercapacitors with tailored characteristics. To attain high capacitance, biochar activation can be performed by the traditional method of mixing of biochar with any strong base followed by baking at high temperature. But this is a time-intensive procedure that is also inefficient in terms of energy consumption. An innovative process for low-temperature activation of biochar was developed by [39]. Their work involved low-temperature (<150°C) plasma treatment to activate yellow pine biochar. This oxygen plasma activation enhanced the biochar microstructure and significantly enhanced the specific capacitance as compared to the chemically activated biochar material. Due to the greater surface area, this results in the creation of widely scattered pore structures.

Agricultural waste like millet straw was utilized to prepare supercapacitor electrode material by thermal modification in the presence of 5:1 ratio of KOH and carbon. Cyclic voltammetry and ESR studies revealed specific capacitance of this activated biochar was larger with high BET specific surface area [27].

3.5.2 Biochar-Based Supercapacitors from Waste Biomass

As a result of population growth and economic activity, the increased generation and improper disposal of waste biomass has received more attention than in the past. Traditional dumping methods may pollute water and soil, causing environmental problems. The viability of producing biochar from waste biomass, which can be used in vivid applications, including the recently researched subject of supercapacitor electrode materials, must be investigated (Table 3.3).

Remediation of water and wastewater by using bio adsorbents generates heavy metal containing spent biochar. Its regeneration and disposal are very challenging, and are energy and cost intensive. Y. Wang *et al.* [41] obtained the biochar from the waste biomass of dairy manure and sewage and converted this biochar loaded with heavy metals into supercapacitor electrodes. Chemical activation was done by Ni adsorption followed by microwave treatment. In comparison to the original biochar supercapacitors, the specific capacitance of biochar supercapacitors improved after Ni loading. After microwave treatment of this hybrid material, the capacitance was further increased by more than 2 times. XRD and XPS studies reflected the conversion of Ni into NiO and NiOOH after microwave treatment which enhanced the capacitance of modified biochar. It also showed good stability of specific capacitance with a loss of less than 2% even after 1,000 cycles.

Biochar can also be obtained from the feedstock waste obtained after the thermochemical process of bio-oil production and it can be utilized for the preparation of the hierarchical carbon having high pore volume and specific surface area. After chemical activation by HNO_3 treatment, specific capacitance of biochar material was found to be improved. It is interesting to note that these hierarchical carbons outperform bio-derived activated carbons, well-ordered mesoporous carbons, and commercially available graphene in terms of capacitive performance. As a result, this approach increases the economic feasibility of thermochemical biofuel processes by converting biochar to a high-value-added carbon material [42].

Organic waste obtained from poultry litter was used to derive biochar through the process of pyrolysis followed by the process of chemical activation. This resulted in a new hierarchically porous super-activated carbon having specific surface area above 3000 $m^2.g^{-1}$ and morphology like graphene. Minor concentration of Phosphorus and Sulphur present in this biochar-based activated carbon distinguishes it from the others, obtained from plant sources. Pontiroli *et al.* 2019, prepared a high-performance symmetric supercapacitor from this functional material which exhibited

Table 3.3 Waste biomass as biochar precursor for supercapacitor application [7, 12, 40, 44–46].

Biomass	Activation process	Surface area SBET ($m^2 \cdot g^{-1}$)	Specific capacitance ($F \cdot g^{-1}$)	Current density	Electrolyte used
Coconut shell	$FeCl_3$; $ZnCl_2$ Pyrolysis, Steam	1874	268	1 A g^{-1}	KOH
Garlic Skin	KOH	2818	427	5 mA.s^{-1}	KOH
Coconut fibers	KOH	2898	142	0.5 A g^{-1}	KOH/ EMIMBF4
Mango stone	$ZnCl_2$ & KOH	1497.8	358.8	0.5 A g^{-1}	
Walnut Shell	Hydrothermal K_2CO_3 sol.	62	255	0.5 A g^{-1}	PVA/KOH
Litchi Shell	KOH	1122.6	222	0.1 A g^{-1}	KOH
Chestnut Shell	Melamine	961	402.8	0.5 A g^{-1}	KOH
Coffee ground	Pyrolysis KOH		404	0.5 A g^{-1}	EMIMTFSI
Almond Shell	KOH/HNO_3	1363.1	283	1 A g^{-1}	KOH
Corncob	KOH	1210	120	1 A g^{-1}	KOH
Sugarcane bagasse	KOH, $CaCl_2$	1982	142.1	0.5 A g^{-1}	KOH

sufficiently high values of specific capacitance, current density, power density and reliability without using any conducting additives. Furthermore, these excellent results were achieved utilizing simple, environmentally friendly electrolytes such as aqueous solutions of KOH and Na_2SO_4. The low cost of the precursor materials, availability and biocompatibility combined with the low environmental effect of electrolyte, indicates a huge

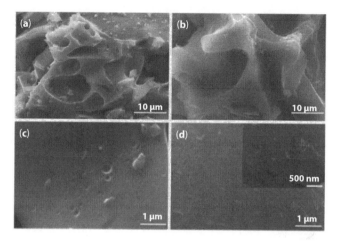

Figure 3.7 SEM Images showing surface morphology of carbonized leather waste biochar (a, c) and activated leather waste biochar (b, d) [47]. Reprinted with the permission of IOP Publishing.

potential for large-scale applications for such devices in transportation, renewable energy grids, and in biomedicine [43].

Studies conducted by Martinez-Casillas *et al.* [47] indicated that footwear leather waste can also be used to generate activated biochar for energy storage devices. This can enhance supercapacitor cell performance as compared to commercially available carbons (Figure 3.7). They derived biochar by pyrolysis of footwear leather wastes at 700°C followed by chemical activation with KOH. This electrode material showed maximum specific capacitance in H_2SO_4 electrolyte. The prepared SC cell displayed high stability with a capacitance of 52 $F.g^{-1}$ at 0.5 mA g^{-1} and loss of only 8% of capacitance even after 5,000 charge/discharge cycles.

3.5.3 Carbon-Based Supercapacitors from Other Methods

To generate high nitrogen biochar from biomass pyrolysis generally low temperature is used but such biochar usually possesses poor conductivity and is not suitable for any energy storage application. W. Zhang *et al.* [48] successfully prepared a biochar-based material with high nitrogen content with conductivity higher than sodium lignin sulfonate, graphene oxide and p-phenylenediamine. Sodium lignin sulfonate served as source of biochar and graphene oxide after reduction provided high conductivity. The biochar electrode possesses greater gravimetric as well as volumetric specific capacitance together with suitable cycle stability in 1M H_2SO_4 electrolyte.

This novel approach paves the way for biochar-based carbon materials for their application as high-performance supercapacitors in energy storage applications.

Carbon nano fibers were synthesized electrochemically by reduction of chloroform in controlled conditions at room temperature and *in situ* functionalization of carbon fibers. Super capacitive behavior of this carbonaceous material was studied (Figure 3.8). Surface studies revealed extensive presence of functionalized species, i.e., -OH, =O, -COOH and amorphous as well as graphitic phase of carbon. Specific capacitance (28 F.g^{-1}) with surface area of 78 m^2.g^{-1} was measured. Thermal treatment caused the increase of surface area to 592 m^2.g^{-1} with no apparent change in the surface morphology and capacitive performance [18].

Hierarchical porous biochar with sulfur and nitrogen co-doping was prepared using mantis shrimp-shell which involved self-activation at 750°C temperature. This material was found to have a distinctive dendritic and uniform surface structure containing regularly interconnected micropores, mesopores and macropores. Electrochemical analysis demonstrated the good specific capacitance of 201 F. g^{-1} at 1 A. g^{-1} in KOH electrolyte. The surface area was found to be 401 m^2.g^{-1} with very high nitrogen (8.2 wt. %) and sulfur (1.16 wt. %). This is a type of in situ template approach from natural biomass which has displayed huge potential for preparing the active electrode materials of supercapacitors [49].

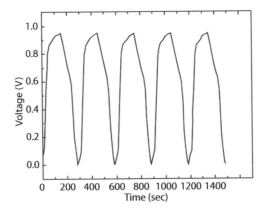

Figure 3.8 Charge-discharge curve of CF electrode [18]. Reprinted with the permission of Elsevier.

3.6 Current Challenges and Future Potential of Biochar-Based Supercapacitor

Electrical energy storage could help to decarbonize the electricity sector by providing a new, carbon-free source leading to the operational flexibility and increased utilization of generation assets, and facilitating the integration of different renewable energy sources. However, the future cost of energy storage technologies is unpredictable, and the value they can provide to the system is contingent on a variety of circumstances. Furthermore, the marginal benefit of storage diminishes when greater energy storage capacity is required. The impact of increasing energy storage capacity on power system operations and generation capacity investments must be investigated. Energy storage offers value by increasing the cost-effectiveness of renewable energy, maximizing the utilization of all installed capacity and lowering total investments in nuclear and gas-fired peaking units. Significant cost reductions in battery storage are necessary to allow large-scale employment. For decarbonization methods which depend on solar and wind energy, energy storage is a necessity, but it can be avoided for low carbon power sources and their varied mixture, e.g., flexible nuclear power [14]. The area of renewable energy not only addresses the issue of carbon reduction but also deals with a variety of different policy goals for example land usage, energy security, employment generation, waste management, etc. In general, it is believed that conflicting policy aims have hampered the growth of biomass utilization. The major environmental concern is related to the alarming concentrations of atmospheric carbon while the other factors like energy supply, waste, and other problems only make a minor contribution. Policymakers must understand that employing limited biomass resources cannot be a solution to battle atmospheric carbon and it is also not efficient or effective. Awareness needs to be raised among the other stakeholders as well, like farmers, industrial communities and societies, that processing their unused biomass resources may yield a better profit [29].

Developing future energy storage systems to increase the use of sustainable electricity like wind or solar in rural regions, as well as improving the use of electric autos, requires breakthroughs in supercapacitor technology. Finally, carbon materials used to fabricate electrodes should have a hierarchical mesoporous structure with suitable pore size distribution greater surface area and conductivity along with term cyclability to attain highly stable and reversible energy storage capacity with high-power density in supercapacitors.

It is sustainable and ecofriendly to obtain hierarchical carbon materials from biological precursors and recycled materials with scalable processes for the large-scale application of supercapacitor in energy storage and conservation. As compared to graphene and metal carbides, biochar feedstock is more economic for the preparation of activated carbon material. Additionally, the performance of electrode materials in supercapacitor can also be tested by well-established commercial current collectors, binders, conducting additives, separators and electrolytes. As a result, the high capacitance performance can be readily achieved in shorter duration in industrial environment [42].

Over the years, different energy storage solutions have been developed and utilized to overcome the unstable nature of renewable, alternative, and clean energy sources like wind, water, and solar energy. Supercapacitors have gained a lot of research attention in comparison to commercially employed lithium-ion batteries because of their higher power density and safer usage due to the prevalent usage of aqueous electrolytes. Biochar (BC) formed from numerous sources such as wood and plant residues, as well as biowastes, has a porous structure and contains N and O components, making it suitable for usage in supercapacitors [26, 38, 47].

Biochar supercapacitors are promising technology for supercapacitors for a great variety of energy and environmental applications due to their superior surface and electrochemical properties [5]. Furthermore, bio-waste-derived high-performance and stable materials, combined with a simple chemical activation process, are definitely a feasible option for generating economically affordable electrodes [37]. Accessible mesopores, which are created by entanglement and the central canal, are required in the best materials. Carbon structures must have mesopores to release a large amount of energy at a fast rate. Mesopores are created through activation mechanisms that include the addition of surface-active groups. In this circumstance, oxygenated groups' pseudocapacitance redox processes also cause an increase in capacitance. In the electrochemical double-layer capacitors, the specific surface area (SSA) of the BC-derived carbon materials also plays a significant role. It makes the active sites available for the absorption and desorption of ionic electrolyte. This also helps in developing high SSA. Both these factors together provide channel for the quick transmission of electrolyte ions and hence increase the conductivity. An interesting future research path is the synthesis of biomass-derived carbon compounds with an interconnected and suitable pore structure. Meanwhile, introducing the heteroatom functional groups of N, P, and O atoms on the surface of biochar can improve both the wettability and capacitance of the electrode material. It can significantly increase their pseudocapacitance capabilities also. Further studies are required on maintaining and improving the cycle stability of biomass-derived carbon-based supercapacitors [12].

Apart from the biochar-based materials, some other carbon materials like multi-walled carbon nanotubes, graphene was also studied for energy storage applications of the supercapacitors. Carbon nanotubes have also been researched extensively due to their superior electrochemical characteristics as supercapacitor electrodes. However, because of their comparatively small surface area (usually less than 500 m^{2}g^{-1}), they have a poor energy density, which limits their practical use [31].

Multi-walled carbon nanotubes (MWNTs) can also be used for accumulation of charge due to their moderate surface area (Figure 3.9). Through the formation of micropores, activation of MWNTs should allow for larger capacitance values. Most of the surface area of carbon materials is made up of micropores, which are incapable of maintaining an electrical double layer. Ion migration through the pores allows ions to partially reach the surface, resulting in an increase in electrolyte resistance. As a result, only low frequencies or a DC technique can be used to extract the stored energy. The performance of capacitor devices is governed by the kinetics of the pseudo-faradaic reactions in three dimensions depending upon the activation energy of charge transfer processes. As a result, such capacitors' rate capability is limited. The capacitance of a real capacitor is always twice that of single electrode materials [50].

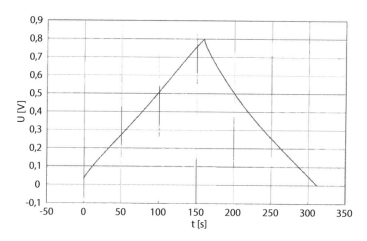

Figure 3.9 Galvanostatic charge/discharge of a supercapacitor built from carbon nanotubes obtained at 7008C and modified by 69% nitric acid:152 mA, 6 M KOH [19]. Reprinted with the permission of Elsevier.

Similarly, graphene can be utilized to prepare supercapacitor electrodes by depositing it on glass and flexible substrates with very low surface resistance, and it can be technology of future in the field of optoelectronics. It can replace the brittle and expensive electrode material such as indium tin oxide (ITO) [21]. Graphene is an appealing material for supercapacitor applications due to its unique electrical characteristics and theoretical surface area (2620 m^2g^{-1}). Due to aggregation of graphene sheets, the surface area of graphene is generally significantly lower than the theoretical value (normally 500 m^2g^{-1}), resulting in lower capacitance values than expected (lesser than 200 $F.g^{-1}$ in aqueous electrolyte, lesser than 120 $F.g^{-1}$ for organic electrolyte, and lesser than 80 $F.g^{-1}$ in ionic liquids) [51].

Modification of biochar-based material provides another approach to introduce desirable changes in the surface structure and functional groups. This can be done by different methods like self-doping, physical activation by modifying the extent and duration of pyrolysis temperature and by employing suitable chemical reagent for chemical modification of surface functional groups (Figure 3.10). It has been demonstrated that different mechanisms of carbon alteration result in samples with distinct chemical properties. As evidenced by data obtained from FTIR, sodium capacity, pH titration, and zeta potential studies, oxidation resulted in the fixing of weakly acidic functional groups. However, the oxidation process created some by-products in the form of humic compounds, which were removed using a sodium hydroxide solution. As indicated by the pore size distribution data, humic compounds tend to get accumulated in micropores, resulting in a loss of microporous structure. Elemental analysis showed presence of Nitrogen in the oxidized samples which was due to the nitrate ions stuck in the pores of the oxidized carbon material. The elimination of acidic functional groups that become unstable at high temperatures can be attributed to hydrogen treatment at elevated temperatures producing a sample with certain basic properties [20].

Self-activation and self-doping are effective methods to improve the electrochemical capacitance by modifying the carbon structure. It can produce high-value biochar from biomass utilization for supercapacitor application. This in situ template approach using naturally abundant biomass as a precursor held a lot of promise for supercapacitor active electrode materials [49]. Studies have also been done on the activation of biochar by plasma, which demonstrated the ability of oxygen plasma to efficiently create porous structures including micropores, mesopores, and macropores of different sizes. This study also reflected that a large surface area together with a proper pore-size distribution can be used to achieve higher specific capacitance [39].

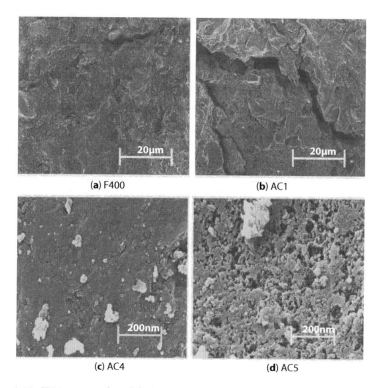

Figure 3.10 SEM images of modified carbon samples, Source: [20] (a) F400 (coal-based commercial granular activated carbon), (b) AC1 (HNO_3 oxidation, H_2O wash, 363K, 9 Hrs.), (c) AC4 (Annealing of F400, 1173K, 3 Hrs.), (d) AC5 ($HNO_3/H_2SO_4/NH_3$ $Na_2S_2O_4/$ $(CH_3CO)_2O$, 298K, 48 Hrs.). Reprinted with the permission of Elsevier.

Another promising application for the future modern grid is the high-temperature biochar supercapacitor, an energy-storage device with high energy and power density. Significant amount of research data is available to showcase that on increasing the temperature, various ionic liquids show significant increase in their viscosities, electrical conductivities, and diffusion coefficients (Figure 3.11). These enhancements are beneficial for enhancing the carbon/electrolyte interface, lowering the resistance associated with redox reactions occurring on carbon surfaces, and facilitating the ion mass transports in the porous structure of carbon materials. The pseudocapacitance associated with redox reactions of oxygen groups on carbon surfaces and electrochemical double layer capacitance can boost the capacity of a supercapacitor. Supercapacitors with a high temperature could store both thermal and electrical energy at the same time [36].

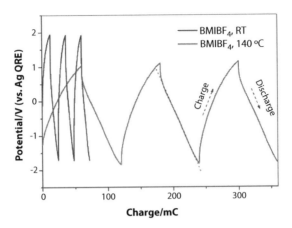

Figure 3.11 Changes of the high temperature biochar electrode as a function of charge passed during three successive constant-current charges and discharges at 0.2 A. g^{-1} and room temperature (black line); and at 1.4 A. g^{-1} and 140°C (red line) [36]. Reprinted under creative commons license.

Even after the tremendous progress in the field of biochar-based materials and their applications for different energy storage and conversion technologies, some existing key concerns still need the attention of future research [7].

1. Surface oxidation, amination, and sulfonation are common methods for biochar functionalization, although they frequently involve complex procedures or harmful and toxic chemicals.
2. It is challenging to keep the conversion efficiency of biomass into biochar without affecting the biochar's subsequent treatments like the tuning of pores and surface chemistry together with the loading of metals and oxides.
3. There is a need to develop a balanced strategy to overcome the harmful effect of contaminants in biochar material on the final performance of supercapacitor electrode because the predicted carbon structure may not necessarily have a good impact on the final performance characteristics.
4. Although the surface characteristics of the biochar may be easily modified, there is no strong flexibility or the capacity to be composited with other different functional materials for additional performance improvement.

Several strategies can be developed and studied in future to find out an amicable and feasible solution of the issues related to the applications of biochar-based supercapacitor electrodes. More work should be put into customizing the precursor biomass or biochar raw material. Chemical structure of plant biomass could be modified with the help of biotechnology so as to enrich the biomass material in certain elements when it is pyrolyzed, and the resulting biochar with certain functions or heteroatom doping can be obtained immediately. This can be a more environmentally friendly technique of biochar functionalization as compared to the present traditional approaches. A cost-effective and readily scaled-up method can be developed to increase the conversion efficiency of biomass to biochar with respect to certain specific properties such as a surface porosity, presence of certain surface functional groups, and an effective surface structure. Carefully selecting and pre-treating biomass before turning it to biochar could be a possible low-cost approach to attain this goal. Theoretical models for the tailor-made biochar material's design and synthesis can be developed. Emphasis should be given to the surface chemistry and molecular interactions so as to reduce the negative impacts of the presence of impurities on biochar's carbon structure and its performance in energy conversion and storage.

To build high-performance biochar-based composites suitable for modern energy applications, components with acceptable physicochemical stability and compatibility with carbon must be chosen. To do this, the components must make full use of biochar's large surface area and retain the structure and porosity of both the components and the biochar. Significant obstacles can be observed while increasing the use of biochar in energy applications; however, the current achievements are both inspirational and comprehensive. If continuous research contributions are produced in this field, there may be great chances to implement practical usage of biochar materials in the domains of renewable energy conversion and storage.

3.7 Conclusion

Carbon materials have the potential to be the most adaptable substrate material in modern energy storage and conversion. Being a carbon-negative process, it will not only aid in the mitigation of serious environmental issues like global warming and pollution but also provide a long-term solution for the storage and conversion of renewable, non-perennial, and dispersive, energy sources such as the sun, wind, water, geothermal, or biomass [52]. Wood, plant and agricultural residues, industrial waste biomass, and sewage, are all promising material for biochar production. Some good examples are red cedar wood, maple wood, pinecones, wooden waste of agriculture and industry, ash tree residue, pomelo peel, millet straw, waste biomass of dairy manure, and sewage, feedstock waste, organic waste obtained from poultry liter, footwear leather waste, etc. Other carbon materials like multi-walled carbon nanotubes, and graphene can also be utilized for supercapacitor applications.

There are two key concerns regarding carbon materials' energy applications. The first difficulty is the process for mass-producing acceptable carbon materials, which determines the material's electrical and electrochemical properties, cost, and environmental impact over its lifetime. Coal and petrochemical-derived carbon compounds are often energy-intensive to produce or involve difficult synthesis conditions. The development of efficient ways for producing carbon materials from renewable resources with excellent performance and minimal environmental impact is a top priority. Biochar, a bio-carbon with abundant surface functional groups and a porosity that can be easily controlled, could be a good contender as a sustainable carbon material in this regard [53].

The second point to consider is the efficiency and performance of carbon materials in a variety of physical situations for various applications. Biochar-based materials have recently been shown to have the potential for energy storage and conversion. Biochar-based materials perform well in supercapacitors, with high specific capacitance and cycle stability. The capacitive properties of biochar-based materials are influenced by a variety of parameters, including surface area, pore characteristics (pore size and distribution), electrical conductivity, and surface interactions. Surface doping with heteroatoms like N, P, and S, as well as surface combining with metal oxide nanostructures, can improve capacitive performance by introducing pseudocapacitive effects. Biochar-based materials perform well in supercapacitors, with high specific capacitance and good cycle stability. The capacitive properties of biochar-based materials are influenced by a variety of parameters, including surface area and pore characteristics (pore size and distribution), as well as electrical conductivity and surface functions. Heteroatom (N, P, S) doping of the surface and its combination with metal oxide nanostructures, can improve capacitive performance by introducing pseudocapacitive effects. In addition to this, specific physical and chemical surface morphologies impart great stability to the electrode materials during charge-discharge cycles. There are many obstacles to overcome while expanding the use of biochar in energy applications, but the existing achievements are both inspirational and comprehensive. It is likely to stimulate more research, allowing biochar-based materials to be employed effectively in modern sustainable energy storage and conversion applications.

References

1. World Energy Outlook 2021, International Energy Agency, Paris, 2021.
2. J. Romero Agüero, E. Takayesu, D. Novosel, and R. Masiello, Grid modernization: challenges and opportunities, Electricity Journal, 30(4), 1–6, 2017.
3. L. Yao, B. Yang, H. Cui, J. Zhuang, J. Ye, and J. Xue, Challenges and progresses of energy storage technology and its application in power systems, Journal of Modern Power Systems and Clean Energy, 4(4), 519–528, 2016.
4. Z. Yang *et al.*, Electrochemical energy storage for green grid, Chemical Reviews, 111(5), 3577–3613, 2011.
5. J. Jiang *et al.*, Highly ordered macroporous woody biochar with ultra-high carbon content as supercapacitor electrodes, Electrochimica Acta, 113, 481–489, 2013.
6. X. Li and S. Wang, Energy management and operational control methods for grid battery energy storage systems, CSEE Journal of Power and Energy Systems, 7(5), 1026–1040, 2021.
7. W. J. Liu, H. Jiang, and H. Q. Yu, Emerging applications of biochar-based materials for energy storage and conversion, Energy and Environmental Science, 12(6), 1751–1779, 2019.
8. A. K. Rohit, K. P. Devi, and S. Rangnekar, An overview of energy storage and its importance in Indian renewable energy sector: Part I – Technologies and Comparison, Journal of Energy Storage, 13, 10–23, 2017.
9. O. Palizban and K. Kauhaniemi, Energy storage systems in modern grids—Matrix of technologies and applications, Journal of Energy Storage, 6, 248–259, 2016.
10. R. M. A., M. S. Palizban Omid, Proceedings of IEEE International Conference on Control System, Computing and Engineering, ICCSCE-2011, Penang, Malaysia, 601, 2011.
11. S. Sabihuddin, A. E. Kiprakis, and M. Mueller, A numerical and graphical review of energy storage technologies, Energies, 8(1), 172–216, 2015.
12. X. Li, J. Zhang, B. Liu, and Z. Su, A critical review on the application and recent developments of post-modified biochar in supercapacitors, Journal of Cleaner Production, 310, 2021.
13. A. K. Rohit and S. Rangnekar, An overview of energy storage and its importance in Indian renewable energy sector: Part II – energy storage applications, benefits and market potential, Journal of Energy Storage, 13, 447–456, 2017.
14. F. J. de Sisternes, J. D. Jenkins, and A. Botterud, The value of energy storage in decarbonizing the electricity sector, Applied Energy, 175, 368–379, 2016.

15. H. Wang, S. Casalongue, Y. Liang, and H. Dai, Ni(OH) 2 Nanoplates Grown on Graphene as Advanced Electrochemical Pseudocapacitor Materials, J. Am. Chem. Soc., 132(21), 7472-7477, 2010.
16. L. Zhang, J. Jiang, N. Holm, and F. Chen, Mini-chunk biochar supercapacitors, Journal of Applied Electrochemistry, 44 (10), 1145–1151, 2014.
17. S. T. Mayer, R. W. Pekala, and J. L. Kaschmitter, The Aerocapacitor: An Electrochemical Double-Layer Energy-Storage Device, Journal of Electrochemical Society, 140 (2), 446–451, 1993.
18. P. v Adhyapak, T. Maddanimath, S. Pethkar, A. J. Chandwadkar, Y. S. Negi, and K. Vijayamohanan, Application of electrochemically prepared carbon nanofibers in supercapacitors, Journal of Power Sources, 109, 105–110, 2002.
19. E. Frackowiak and F. Beguin, Carbon materials for the electrochemical storage of energy in capacitors, Carbon, 39, 937-950, 2001.
20. P. Chingombe, B. Saha, and R. J. Wakeman, Surface modification and characterisation of a coal-based activated carbon, Carbon N Y, 43 (15), 3132–3143, 2005.
21. R. Garg, S. Elmas, T. Nann, and M. R. Andersson, Deposition Methods of Graphene as Electrode Material for Organic Solar Cells, Advanced Energy Materials, 7(10), 2017.
22. A. Szabó, C. Perri, A. Csató, G. Giordano, D. Vuono, and J. B. Nagy, Synthesis methods of carbon nanotubes and related materials, Materials, 3(5), 3092–3140, 2010.
23. A. R. Karaeva, N. v Kazennov, E. A. Zhukova, and V. Z. Mordkovich, Carbon nanotubes by continuous growth, pulling and harvesting into big spools, Material Today: Proceedings, 5, 25951-25955, 2018.
24. J. J. Manyà, Pyrolysis for biochar purposes: A review to establish current knowledge gaps and research needs, Environmental Science and Technology, 46 (15), 7939–7954, 2012.
25. W. J. Liu, H. Jiang, and H. Q. Yu, Development of Biochar-Based Functional Materials: Toward a Sustainable Platform Carbon Material, Chemical Reviews, 115(22), 12251–12285, 2015.
26. W. J. Liu, H. Jiang, and H. Q. Yu, Development of Biochar-Based Functional Materials: Toward a Sustainable Platform Carbon Material, Chemical Reviews, 115(22), 12251–12285, 2015.
27. Y. Ding, T. Wang, D. Dong, and Y. Zhang, Using Biochar and Coal as the Electrode Material for Supercapacitor Applications, Frontiers in Energy Research, 7, 2020.
28. D. Woolf, J. E. Amonette, F. A. Street-Perrott, J. Lehmann, and S. Joseph, Sustainable biochar to mitigate global climate change, Nature Communications, 1(5), 2010.
29. M. Fowles, Black carbon sequestration as an alternative to bioenergy, Biomass and Bioenergy, 31 (6), 426–432, 2007.
30. Q. Zhang, E. Uchaker, S. L. Candelaria, and G. Cao, Nanomaterials for energy conversion and storage, Chemical Society Reviews, 42 (7), 3127–3171, 2013.

31. M. Sevilla and R. Mokaya, Energy storage applications of activated carbons: Supercapacitors and hydrogen storage, Energy and Environmental Science, 7(4), 1250–1280, 2014.
32. Y. Lin, H. Zhao, F. Yu, and J. Yang, Design of an extended experiment with electrical double layer capacitors: Electrochemical energy storage devices in green chemistry, Sustainability (Switzerland), 10 (10), 2018.
33. J. Wang, C. Ma, L. Su, L. Gong, D. Dong, and Z. Wu, Self-Assembly/Sacrificial Synthesis of Highly Capacitive Hierarchical Porous Carbon from Longan Pulp Biomass, ChemElectroChem, 7(22), 4606–4613, 2020.
34. M. Genovese and K. Lian, Polyoxometalate modified pinecone biochar carbon for supercapacitor electrodes, Journal of Materials Chemistry A, 5(8), 3939–3947, 2017.
35. C. Wan, Y. Jiao, and J. Li, Core-shell composite of wood-derived biochar supported MnO_2 nanosheets for supercapacitor applications, RSC Advances, 6(69), 64811–64817, 2016.
36. J. Jiang, High Temperature Monolithic Biochar Supercapacitor Using Ionic Liquid Electrolyte, Journal of the Electrochemical Society, 164 (8), H5043–H5048, 2017.
37. L. Kouchachvili and E. Entchev, Ag/Biochar composite for supercapacitor electrodes, Materials Today Energy, 6, 136–145, 2017.
38. Y. Li et al., Super-hydrophilic microporous biochar from biowaste for supercapacitor application, Applied Surface Science, 561, 2021.
39. R. K. Gupta, M. Dubey, P. Kharel, Z. Gu, and Q. H. Fan, Biochar activated by oxygen plasma for supercapacitors, Journal of Power Sources, 274, 1300–1305, 2015.
40. C. Senthil and C. W. Lee, Biomass-derived biochar materials as sustainable energy sources for electrochemical energy storage devices, Renewable and Sustainable Energy Reviews, 137, 110464, 2021.
41. Y. Wang *et al.*, Converting Ni-loaded biochars into supercapacitors: Implication on the reuse of exhausted carbonaceous sorbents, Scientific Reports, 7, 2017.
42. H. Jin, X. Wang, Z. Gu, and J. Polin, Carbon materials from high ash biochar for supercapacitor and improvement of capacitance with HNO_3 surface oxidation, Journal of Power Sources, 236, 285–292, 2013.
43. D. Pontiroli et al., Super-activated biochar from poultry litter for high-performance supercapacitors, Microporous and Mesoporous Materials, 285, 161–169, 2019.
44. H. Fu *et al.*, Walnut shell-derived hierarchical porous carbon with high performances for electrocatalytic hydrogen evolution and symmetry supercapacitors, International Journal of Hydrogen Energy, 45 (1), 443–451, 2020.
45. J. Li, Q. Jiang, L. Wei, L. Zhong, and X. Wang, Simple and scalable synthesis of hierarchical porous carbon derived from cornstalk without pith for high capacitance and energy density, Journal of Materials Chemistry A, 8(3), 1469–1479, 2020.

46. J. Li, L. Wei, Q. Jiang, C. Liu, L. Zhong, and X. Wang, Salt-template assisted synthesis of cornstalk derived hierarchical porous carbon with excellent supercapacitance, Industrial Crops and Products, 154, 112666, 2020.
47. D. C. Martínez-Casillas, I. L. Alonso-Lemus, I. Mascorro-Gutiérrez, and A. K. Cuentas-Gallegos, Leather Waste-Derived Biochar with High Performance for Supercapacitors, Journal of the Electrochemical Society, 165(10), A2061–A2068, 2018.
48. W. Zhang, Y. Zou, C. Yu, and W. Zhong, Nitrogen-enriched compact biochar-based electrode materials for supercapacitors with ultrahigh volumetric performance, Journal of Power Sources, 439, 2019.
49. S. Huang, Y. Ding, Y. Li, X. Han, B. Xing, and S. Wang, Nitrogen and Sulfur Co-doped Hierarchical Porous Biochar Derived from the Pyrolysis of Mantis Shrimp Shell for Supercapacitor Electrodes, Energy and Fuels, 35(2), 1557–1566, 2021.
50. E. Frackowiak and F. Beguin, Carbon materials for the electrochemical storage of energy in capacitors, 2001.
51. Y. Zhu *et al.*, Carbon-Based Supercapacitors Produced by Activation of Graphene, Science, 332, 1537, 2011.
52. Y. Lin, F. Li, Q. Zhang, G. Liu, and C. Xue, Controllable preparation of green biochar based high-performance supercapacitors, Ionics (Kiel), 28 (6), 2525–2561, 2022.
53. Z. Li, D. Guo, Y. Liu, H. Wang, and L. Wang, Recent advances and challenges in biomass-derived porous carbon nanomaterials for supercapacitors, Chemical Engineering Journal, 397, 125418, 2020.

4
Energy Storage Units for Frequency Management in Nuclear Generators-Based Power System

Boopathi D.[1]*, Jagatheesan K.[1], Sourav Samanta[2], Anand B.[3] and Satheeshkumar R.[1]

[1]*Paavai Engineering College, Tamil Nadu, India*
[2]*University Institute of Technology, University of Burdwan, West Bengal, India*
[3]*Hindusthan College of Engineering and Technology, Tamil Nadu, India*

Abstract

In this chapter, the Load Frequency Management (LFM) of an interconnected nuclear power network with an energy storage unit (ESU) is explained. The impact of the ESU in the proposed power system during emergency loading is analyzed in detail. A Proportional Integral Derivative (PID) controller is designed and implemented as a secondary controller for the LFM of the proposed power network. The Ant colony optimization (ACO) technique is utilized to optimize the gain parameters of the proposed controller. The impact of various ESUs such as battery energy storage system (BESS), fuel cell (FC), redox flow battery (RFB), ultra-capacitor (UC), proton exchange membrane (PMC) based FC, and supercapacitor energy system (SCES) is analyzed in detail. The proposed optimization technique-based controller is performed well and regulates the stability of the system frequency, in terms of quick settling time, minimal peak over, and undershoots.

Keywords: Energy storage units, load frequency management, ant colony optimization, interconnected power system, frequency deviation

4.1 Introduction

In recent years the development of industries and modernization of domestic life has created more power requirements. To balance the difference

---

Corresponding author: boopathime@gmail.com

Sandeep Dhundhara, Yajvender Pal Verma, and Ashwani Kumar (eds.) *Energy Storage Technologies in Grid Modernization*, (105–134) © 2023 Scrivener Publishing LLC

between power requirement and power generation, countries are moving towards installing new power plants and increasing the standing power plant capacity. While increasing the power generating capacity, due to the system's complex more power quality issues occur during the emergency loading time. The LFM is a scheme to control and maintain the frequency-related issues in the power grid. Many research people are handling these issues by implementing various secondary controllers, optimization techniques to optimize the controller gain parameters, and energy storage units. The purpose of the following literature review is to analyze and find the research gap in the LFM.

Improved salp swarm computation algorithm has been proposed for LFM of microgrid (MG) power system which includes renewable energy sources (wind and solar), and energy storage units like BESS, FC by [1]. The author in [2] studied the impact and performance of the various energy storage systems (ESS) with grid-connected power systems. An advanced controller is developed for LFM of power network which has PMC-based FC in [3]. Multiverse optimizer (MVO) based PID controller has been implemented to perform LFM of an interconnected power system in [4]. UC and RFB play a vital role in [5] the LFM of an isolated hybrid power network. The author in [6] analyzed the impact of the SMES. RFB and HAE with single area nuclear power plant for LFM by developing an ACO – PID controller.

An opposition-based Harmonic Search (OHS) based integral controller was developed by the authors in [7] to perform LFM for interconnected multi-source power network incorporated with RFB. The author in [8] has proposed and analyzed the impact of a hybrid energy storage system (HESS) consisting of supercapacitor and RFB unit by implementing the neural network-based PID controller. In [9] a CO - PID controller to perform LFC of wind power form with BESS and FC is implemented. A multi-area microgrid which consists of RFB and SMES is investigated by [10] for LFM. CES and an SMES incorporated microgrid is examined by PSO – PID controller in [11]. An integral controller is utilized as a secondary controller for single area power system with BESS and supercapacitor in [12]. To perform LFM in a microgrid consisting of FC, Flywheel a fuzzy logic controller (FLC) is implemented by [13].

A CES incorporated power system is examined by hybrid PSO and GA-based PID controller for LFM in [14]. The author [15] investigated a 2-area interconnected power system with CES by FLC. Flower pollination algorithm-based PID controller is designed by [16] to implement LFM in microgrid with CES. A classical PI controller is designed by [17] for LFM of hybrid microgrid with PEM FC. The author in [18] investigated a single and double reheated thermal system with SMES and RFB by stochastic

Table 4.1 Summary of literature review.

Author/Year	Energy storage units	Computation techniques/ Controller utilized	System considered	Ref. no.
Sahu et al., 2018	BESS, FC	Improved Salp swarm/fuzzy PID	Microgrid – Micro turbine + Diesel engine generator	[1]
Akram et al., 2020	ESS	Model optimizer	Single area thermal + wind + PV + energy storage system	[2]
Yildirim, B., 2021	BESS, PEM, FC	Advanced controller	Microgrid – PV+wind+Fuelcell+BESS	[3]
Sharma et al., 2021	RFB	Multiverse optimizer/PID	Six area power system with fuel cell	[4]
Mudi et al., 2021	RFB, UC	Multiverse optimizer/PID	Solar thermal + PV	[5]
Dhanasekaran et al., 2020	RFB, HAE, SMES	Ant Colony Optimization/PID	Single area nuclear	[6]
Shankar et al., 2016	RFB	Opposition-based Harmonic Search – I	Thermal + Hydro + Gas	[7]
Xu et al., 2017	Supercapacitor, RFB	Hammerstein-type neural network/PID	Thermal	[8]
Boopathi et al., 2022	BESS, FC	Ant Colony Optimization/PID	Wind	[9]

(Continued)

Table 4.1 Summary of literature review. (*Continued*)

Author/Year	Energy storage units	Computation techniques/ Controller utilized	System considered	Ref. no.
Sadoudi *et al.*, 2021	RFB, SMES	Elephant Herding Optimization PIDN	Microgrid (Diesel + PV + wind)	[10]
Ali *et al.*, 2018	CES, SMES	Particle Swarm Optimization/PID	Micro hydro power system	[11]
Zhou *et al.*, 2017	Supercapacitor, BESS	Classical Integral	Microgrid with RESs	[12]
Abazari *et al.*, 2019	FC, Flywheel	Fuzzy logic controller	Diesel, wind + energy storage system	[13]
Khadanga and Kumar, 2019	CES	Hybrid Particle Swarm Optimization & Genetic Algorithm/PID	Thermal + Hydro + Gas	[14]
Arya, 2019	CES	Fuzzy Logic Controller	Thermal + Hydro	[15]
Ali *et al.*, 2018	CES	Flower pollination algorithm/PID	Micro hydro	[16]
Sharma and Mishra, 2017	PEM FC	Classical PI	PV	[17]

(*Continued*)

Table 4.1 Summary of literature review. (*Continued*)

Author/Year	Energy storage units	Computation techniques/ Controller utilized	System considered	Ref. no.
Jagatheesan et al., 2016	RFB, SMES	Stochastic Particle Swarm Optimization/PID	Thermal	[18]
Jagatheesan et al., 2018	SMES	Ant Colony Optimization/ PID	Thermal	[19]
Boopathi et al., 2021	FC, BESS	Particle Swarm Optimization/PID	Thermal + PV + wind	[20]
Kumarakrishnan et al., 2020	BESS	Particle Swarm Optimization/PID	Thermal + hydro + gas + nuclear	[21]
Dhundhara and Verma, 2018	CES	Particle Swarm Optimization	Thermal + hydro + gas	[22]
Jin et al., 2016	BESS	-	Thermal + hydro + nuclear	[23]
Kouba et al., 2016	RFB	MVO/PI	Wind + diesel	[24]
Kumar and Sharma, 2020	CES	Whale optimization/PI	Thermal + hydro + gas	[25]

PSO – PID controller. An SMES incorporated power system is examined in [19] with ACO – PID controller. FC, BESS incorporated microgrid power network is investigated by PSO – PID controller by the author in [20]. A PSO – PID controller is developed for the LFM of the single area power system with BESS in [21]. An interconnected multiple sources power network is examined by PSO – PI controller in [22]. In [23] AGC of a real-time implementation of BESS in South Korea was studied. The author in [24] developed a MVO – PI controller for LFC of RFB incorporated power system. A interconnected thermal + hydro + gas power network is examined by whale optimization algorithm tuned PI controller in [25]. The summary of the literature review is shown in Table 4.1.

The literature review helped to find the impact and role of ESUs in the power system for LFM. In recent days, renewable energy sources (RESs) are more penetrating in the power generation sector. While used in the power system, the ESU much needs one, because most of the RESs provide non-linear power output. To regulate that power output, ESU is utilized. Other power systems also use the ESU to compensate for the small load demands. In the literature review, most of the energy storage units the researchers are investigating are in RESs-based power system only. So, in this chapter the research gap is filled by analyzing the energy storage units' impact on nuclear power plants.

4.1.1 Structure of the Chapter

This chapter is structured as per the following plan: section 4.1 provides the complete and detailed analysis of the ESU in the power system by literature review. A proposed Simulink model, different ESU which are analyzed in the chapter is given in section 4.2. The basic structure and role of controllers are explained in section 4.3. Section 4.4 discusses the optimization algorithm implemented to tune the controller. The detailed impact analysis of ESU is given in section 4.5. Results and discussion are in section 4.6.

4.1.2 Objective of the Chapter

- To identify and design an energy storage unit, which can be incorporated with the power system.
- To develop a secondary controller to study the LFM in the power system.
- Analysis of the performance and contribution of the ESU in the power system for LFM.

4.2 Investigated System Modeling

Nuclear power is one of the upcoming electrical power generators. Because of the shortage of fossil fuels, renewable energy sources have become part of electrical power generation; due to the environmental factor the renewable sources are not able to generate the power to overcome the load demand. So large power production may depend on nuclear power plants. In the report issued in January 2020 titled "Optimal Generation Capacity

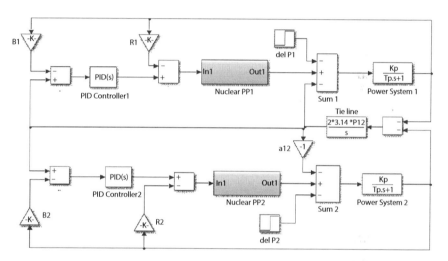

Figure 4.1a Proposed Simulink model of power system.

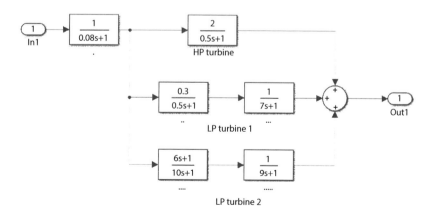

Figure 4.1b Isolated model of nuclear power plant.

Mix for 2029-30", the Central Electricity Authority India planned to install 12,000 MW capacity of nuclear power plants in India. A two-area nuclear power plant is proposed in the chapter to investigate the performance of the system and the impact of the ESU with the proposed power system. The proposed model is shown in Figure 4.1a and, the isolated model of the nuclear power plant is shown in Figure 4.1b [6].

The nuclear power plant static model was designed by the governor, one High Pressure (HP), two parallel Low Pressure (LP) turbines. To improve the performance of the turbine a reheater is connected in between the turbines.

4.2.1 Battery Energy Storage System (BESS) Model

The BESS unit is designed to address the critical loads' instantaneous energy needs in microgrid systems. At the time of the underloading the battery acted as load and gets charged. At the time of overloading the battery acted as source. G_{BESS} in equation 4.2 [3] creates the transfer function provided by the BESS device's transient response. The time constant of BESS is represented by T_{BESS} in equation 4.1. BESS incorporated model is shown in Figure 4.2a and the mathematical expression of BESS is given in Figure 4.2b.

$$G_{BESS} = \frac{1}{1 + T_{BESS}S} \quad (4.1)$$

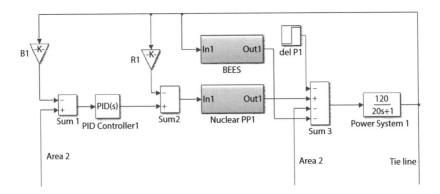

Figure 4.2a Proposed power system model with BESS.

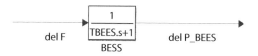

Figure 4.2b Transfer function model of BESS.

4.2.2 Fuel Cell (FC) Model

Some indicators will exceed their standards due to sudden changes in load, reducing their service life. As a result, an energy management unit with an energy storage device is required to improve the fuel cell's dynamic performance, reduce damage caused by rapid load change, and extend service life. Figure 4.3 depicts FC incorporated into the proposed model. FC is supported to balance the power demand as power source.

4.2.3 Redox Flow Battery (RFB) Model

RFB is one of the most visible electrochemical static Energy Storage (ES) devices. The term "redox" refers to the chemical and oxidation reactions that the RFB uses to store energy in the electrolyte solution. During the oxidation process, an electron is discharged from a high potential state to the negative terminal of the battery. To perform useful work, the electron travels through an external circuit. Finally, the electron is accepted by the positive side of the battery. The way new electrolytes are consistently directed into the battery to maintain energy levels gives it an advantage over other ES devices. RFBs contain sulphuric acid (H_2SO_4) and vanadium pentoxide ions (V_2O_5) as an electrolyte [4]. The basic chemical reactions that occur are shown in equations 4.2 and 4.3.

At positive electrode

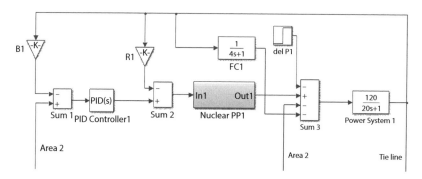

Figure 4.3 Proposed power system model with FC.

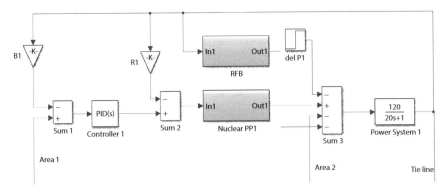

Figure 4.4a Proposed power system model with RFB.

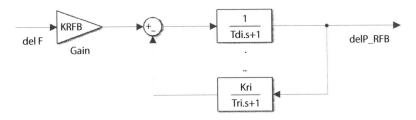

Figure 4.4b Transfer function model of RFB.

$$V^{4+} \leq > V^{5+} + e^- \qquad (4.2)$$

At negative electrode

$$V^{3+} + e^- (=) V^{2+} \qquad (4.3)$$

RFB has a wide range of applications in the frequency regulation problem. When there is a delay in the governor response due to their drowsy reaction, RFBs provide a brisk and fast response and thus suit well for frequency regulation in system. They are suitable for applications with power ratings ranging from tens of KW to tens of MW in [4]. RFB connected model and the RFB transfer function is shown in Figures 4.4a and 4.4b.

4.2.4 Proton Exchange Membrane (PEM) Based FC Model

In this power system, a proton exchange membrane (PEM) based fuel cell is used. PEM fuel cells are the most common fuel-cell type and are widely used in a variety of applications. PEM – FC consists of the fuel cell,

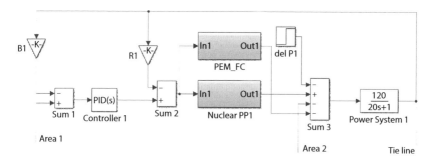

Figure 4.5a Proposed power system model with PEM-FC.

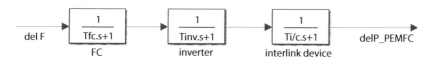

Figure 4.5b Transfer function model of PEM-FC.

inverter, and interlinked device. The incorporation of the PEM-FC with the proposed system and the mathematical expression of the PEM-FC is given in Figures 4.5a and 4.5b, respectively.

4.2.5 Ultra-Capacitor (UC) Model

The capacitor stores electrical energy in the form of electrostatic energy. The amount of energy storage is directly proportional to the voltage applied between the plates (V) charge stored in the capacitor is $Q \propto V$. The capacitor proportional constant is (C) is called capacitance. The charge on a capacitor is given in equation 4.4.

$$Q = C \times V \quad (4.4)$$

Where
Q = Charge (Coulombs)
C = Capacitance (Farads)
V = Voltage (Volts)

Another type of capacitor, an ultra-capacitor (UC), has large conductive plates (electrodes). UC has excellent energy storage capacity because

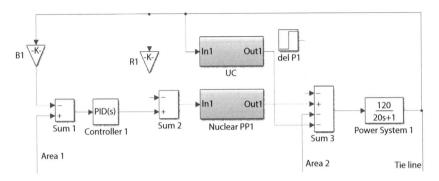

Figure 4.6a Proposed power system model with UC.

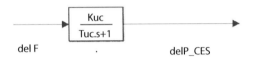

Figure 4.6b Transfer function model of UC.

of high capacitance up to hundreds of farads, a small gap between the electrode and, electrode have high surface area. As previously stated, an ultra-capacitor is an electrochemical device that stores charge electrostatically by utilizing two porous electrodes, typically made of activated carbon immersed in an electrolyte solution. This configuration effectively creates two capacitors, one at each carbon electrode and linked in series.

UC incorporated proposed model is shown in Figure 4.6a and the transfer function model of the UC is given in Figure 4.6b [5].

4.2.6 Supercapacitor Energy Storage (SCES) Model

The SCES unit is an energy storage device that is integrated with a power conversion system that includes a rectifier/inverter, capacitor, and some protective. SCES incorporated system model and transfer function model is shown in Figures 4.7a and 4.7b, respectively. SCES also acted as ultra-capacitor its charges during underloading and discharging during overloading.

4.3 Controller and Cost Function

In the implementation process of the LFC/LFM scheme in the power system, the secondary controller plays a major role. The primary controller

Impact of Energy Storage Systems in Nuclear Power Plants 117

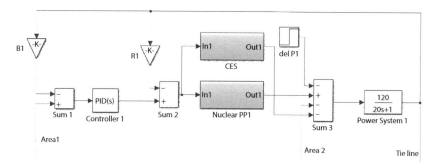

Figure 4.7a Proposed power system model with SCES.

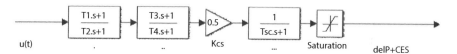

Figure 4.7b Transfer function model of the SCES.

in the power system is not suitable for controlling/maintaining the system stability at sudden loading conditions. In this chapter, the PID controller is designed for implementing LFC/LFM. The PID controller is a commonly used industrial-based controller. It is designed by three separate controllers (Proportional, Integral, Derivative). The mathematical function of the

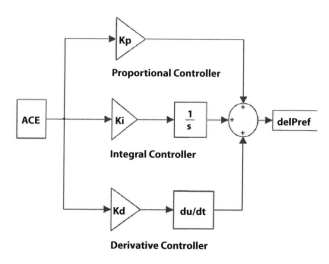

Figure 4.8 Simple structure of PID controller.

PID controller is given in equation 4.5. A simple structure of the PID controller is shown in Figure 4.8.

$$G_{PID} = K_P + \frac{K_I}{S} + S\, K_D \qquad (4.5)$$

The objective function is playing an important role in the controller tuning process for minimizing the error signal and helps to find the fitness value J. In this chapter, the integral time absolute error (ITAE) cost function is involved to get an optimal controller gain parameter. The mathematical expression of the ITAE cost function is given in equation 4.6.

$$J_{ITAE} = \int^{tsim} (delF).t\, dt \qquad (4.6)$$

4.4 Optimization Methodology

Ant colony optimization (ACO) technique is one of the most popular optimization techniques in various fields. The development of the technique was inspired by the food searching process of ants and also the finding of the shortest path between food source and nest. First this technique is

Table 4.2 ACO technique optimized proposed controller gain parameters.

Investigated system/ Gain parameters of the proposed controller	PID controller 1			PID controller 2		
	K_P	K_I	K_D	K_P	K_I	K_D
System without ESU	0.99	1	0.31	0.74	0.87	0.17
System with BESS	1	0.99	0.02	0.99	0.99	0.44
System with CES	0.97	0.98	0.18	0.98	0.71	0.6
System with FC	0.95	0.99	0.29	0.53	0.9	0.19
System with PEM-FC	0.86	1	0.18	0.81	0.93	0.95
System RFB	0.99	0.96	0.02	0.87	1	0.05
System UC	0.96	1	0.29	0.37	0.62	0.48

implemented to solve the traveling salesman problem. By implementing the ACO technique the proposed PID controller gain parameters are optimized. The optimized controller gain parameters are presented in Table 4.2 [6, 19].

4.5 Impact Analysis of Energy Storage Units

The proposed power system Simulink model is investigated with 1% of SLP in the MATLAB 2014a simulating platform. The impact analysis of each ESU is studied in this section in detail as given below:

4.5.1 Impact of BESS

To investigate the impact of the BESS for LFC of the proposed interconnected power system, BESS is incorporated with the proposed system which is shown in section 4.2. The impact of the BESS is analyzed by comparing the system response with/without BESS in the proposed power system. The frequency oscillation comparison of area 1 and area 2 is given in Figures 4.9a and 4.9b, respectively. The numerical values in Figures 4.9a and 4.9b are presented in Table 4.3.

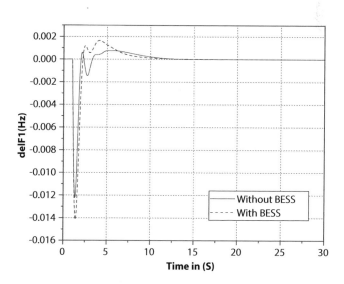

Figure 4.9a delF1 comparison of with/without BESS.

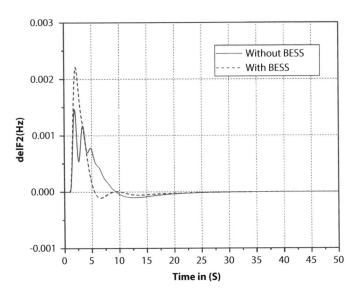

Figure 4.9b delF2 comparison of with/without BESS.

Table 4.3 Time domain-specific parameters of del F1 & F2 of proposed system with/without BESS.

Time domain-specific parameter/ System	del F1			del F2		
	T_S (s)	P_{OS}(Hz)	P_{US}(Hz)	T_S (s)	P_{OS}(Hz)	P_{US}(Hz)
System without BESS	15.2	7.5 × 10^{-4}	0.0122	48	1.46 × 10^{-3}	1 × 10^{-4}
System with BESS	15	1.6 × 10^{-3}	0.0141	35	2.21 × 10^{-3}	1 × 10^{-4}

The impact analysis of the BESS is given in Figures 4.9a and 4.9b and numerical values in Table 4.3 clearly show that the BESS is helping to settle the frequency deviation F1 (15s< 15.2s) & F2 (35s<48s) of the power system at emergency conditions. Where T(s) is settling time of the system frequency, P_{OS} referred as overshoot, and P_{US} is denoted as undershoot.

4.5.2 Impact of FC

To analysis the impact of the FC for the LFM of the proposed power system, an FC is interconnected with the nuclear power plant. The impact analysis of the FC in the proposed system is given in Figures 4.10a and 4.10b, and the nominal values from the graph are reported in Table 4.4.

The impact analysis of the FC is given in Figures 4.10a and 4.10b and numerical values reported in Table 4.4 clearly show that the FC is helping

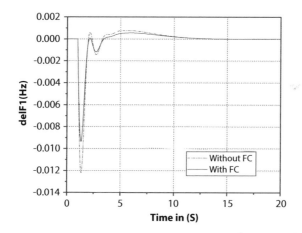

Figure 4.10a delF1 comparison of with/without FC.

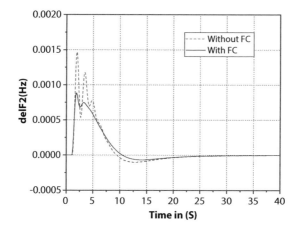

Figure 4.10b delF2 comparison of with/without FC.

Table 4.4 Time domain-specific parameters of del F1&F2 of proposed system with/without FC.

Time domain-specific parameter/System	del F1			del F2		
	T_S (s)	P_{OS}(Hz)	P_{US}(Hz)	T_S (s)	P_{OS}(Hz)	P_{US}(Hz)
System without FC	15.2	7.5×10^{-4}	0.0122	48	1.46×10^{-3}	1×10^{-4}
System with FC	15.2	5.5×10^{-4}	0.009	35	1.4×10^{-3}	1×10^{-4}

to settle the frequency deviation F1 (15.2s = 15.2s) & F2 (35s < 48s) quick in the area 2 and provide the minimal peak values of the system at emergency conditions.

4.5.3 Impact of RFB

To study and analyze the impact of the RFB, an RFB unit is incorporated with the proposed system. The del F is fed as input to the RFB. The frequency deviation of area 1 and area 2 graphically is shown in Figures 4.11a and 4.11b respectively. Numerical values in Figure 4.11 are reported in Table 4.5.

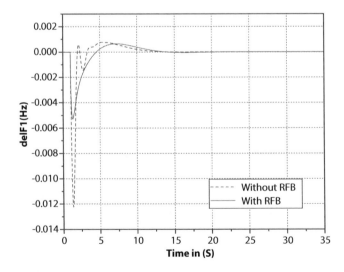

Figure 4.11a delF1 comparison of with/without RFB.

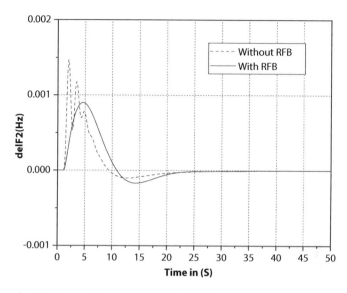

Figure 4.11b delF2 comparison of with/without RFB.

Table 4.5 Time domain-specific parameters of del F1&F2 of the proposed system with/without RFB.

Time domain-specific parameter/ System	del F1			del F2		
	T_S (s)	P_{OS}(Hz)	P_{US}(Hz)	T_S (s)	P_{OS}(Hz)	P_{US}(Hz)
System without RFB	15.2	7.5 × 10^{-4}	0.0122	48	1.46 × 10^{-3}	1 × 10^{-4}
System with RFB	30	6.5 × 10^{-4}	5.2 × 10^{-3}	41	9 × 10^{-4}	1.7 × 10^{-4}

From the response comparison of with/without RFB in the proposed power system in Figures 4.11a and 4.11b, numerical values in Table 4.5 clearly show that the RFB provides the improvement in the frequency deviation area 2 (41s < 48s) and minimize the peak shoots in both areas.

4.5.4 Impact Analysis of the PEM-FC

The performance of the PEM-FC for LFC of interconnected power system is the analysis in this section. PEM-FC is an interconnected with the proposed power system (two-area nuclear) and the system performance is analyzed. The frequency deviation of area 1 and 2 are represented in Figures 4.12a and 4.12b, respectively. The numerical values in Figures 4.11a and b are reported in Table 4.6.

The impact of the PEM-FC for the LFM of the proposed power system is analyzed in detail and reported graphically in Figures 4.12 a and b.

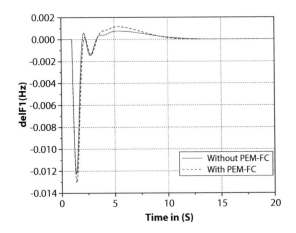

Figure 4.12a delF1 comparison of with/without PEM-FC.

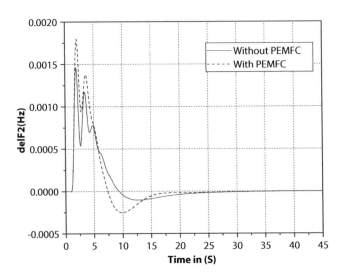

Figure 4.12b delF2 comparison of with/without PEM-FC.

Table 4.6 Time domain-specific parameters of del F1&F2 of proposed system with/without PEM-FC.

Time domain-specific parameter/System	del F1			del F2		
	T_s (s)	P_{OS}(Hz)	P_{US}(Hz)	T_s (s)	P_{OS}(Hz)	P_{US}(Hz)
System without PEM-FC	15.2	7.5 × 10^{-4}	0.0122	48	1.46 × 10^{-3}	1 × 10^{-4}
System with PEM-FC	11	1.5 × 10^{-3}	0.013	35	1.8 × 10^{-3}	2.5 × 10^{-4}

The numerical values of Figures 4.12a and b is in Table 4.6. It is clearly demonstrated that the PEM FC performed well in the proposed system for LFM. It provides a better response in terms of fast settling of the frequency oscillation F1 (11s < 15.2s) & F2 (35s < 48s) and minimal peak values.

4.5.5 Impact Analysis of UC

The impact of the UC in the proposed system for LFC is studied, and the frequency deviation comparison of area 1 and 2 is shown in Figures 4.13a

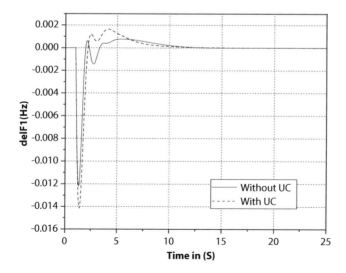

Figure 4.13a delF1 comparison of with/without UC.

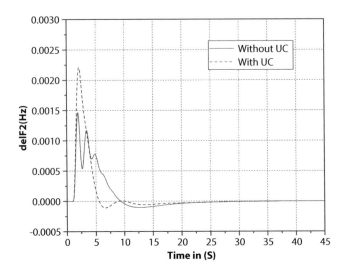

Figure 4.13b delF2 comparison of with/without UC.

Table 4.7 Time domain-specific parameters of del F1&F2 of proposed system with/without UC.

Time domain-specific parameter/ System	del F1			del F2		
	T_s (s)	P_{os}(Hz)	P_{us}(Hz)	T_s (s)	P_{os}(Hz)	P_{us}(Hz)
System without UC	15.2	7.5×10^{-4}	0.0122	48	1.46×10^{-3}	1×10^{-4}
System with UC	15	1.65×10^{-3}	0.014	35	2.2×10^{-3}	1.1×10^{-4}

and 4.13b, respectively. The nominal values in Figures 4.13a and 4.13b are reported in Table 4.7.

The performance of the UC in the proposed system for LFM is clearly demonstrated in Figures 4.13a and 13b. The nominal values of Figure 4.13 are reported in Table 4.7. From Figure 4.13 and Table 4.7, the UC has made a positive impact in the proposed system in terms of quick setting time F1 (15s < 15.2s) & F2 (35s < 48s) and minimal peak values.

4.5.6 Impact Analysis of SCES

To study and analyze the impact of the SCES, a SCES unit is incorporated with the proposed system. The del F is fed as input to the SCES. The frequency deviation of area 1 and 2 are graphically shown in Figures 4.14a

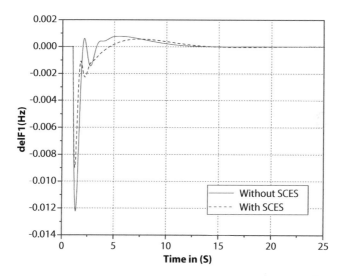

Figure 4.14a delF1 comparison of with/without SCES.

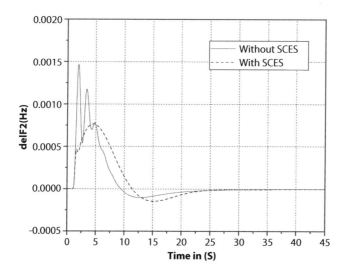

Figure 4.14b delF2 comparison of with/without SCES.

Table 4.8 Time domain-specific parameters of del F1&F2 of proposed system with/without SCES.

Time domain-specific parameter/ System	del F1			del F2		
	T_s (s)	P_{OS}(Hz)	P_{US}(Hz)	T_s (s)	P_{OS}(Hz)	P_{US}(Hz)
System without SCES	15.2	7.5×10^{-4}	0.0122	48	1.46×10^{-3}	1×10^{-4}
System with SCES	29	5.5×10^{-4}	9×10^{-3}	43	7.6×10^{-4}	1.4×10^{-4}

and 4.14b, respectively. Numerical values in Figure 4.14 are reported in Table 4.8.

From Figures 4.14a, 4.14b, and Table 4.8 the SCES has not made a positive impact on the proposed power system for LFC. Because the SCES incorporated model takes more time to settle the oscillation F2 (43s < 48s) in the system frequency at emergency conditions than the system without SECS.

4.6 Result and Discussion

The various energy storage units such as BESS, FC, RFB, PEM-FC, UC, and SCES are incorporated with the proposed two-area nuclear power system to analyze the impact of each ESU to implement the LFM scheme to maintain the system stability and frequency during emergency loading conditions. The comparative analysis of del F1, del F2, and tie-line power deviation is plotted in the 4.15a – 4.15c, respectively. From section 4.5 and Figure 4.15, the impact of the ESU is clearly shown, the PEM-FC performed well over other ESUs. PEM-FC helps to minimize the peak shoot and bring back the frequency to the normal state from oscillation.

From the frequency deviation comparison of the proposed power system incorporated with various ESU in Figures 4.15a, 4.15b and tie-line power flow comparison in Figure 4.15c it is strongly demonstrated that energy storage units create some impact in the system for LFM. The different ESUs are provided variety of performance such as delF1 for BESS (15s < 15.2s), FC (15.2s = 15.2s), RFB (30s > 15.2s), PEMFC (11s <15.2s), UC (15s< 15.2s), and SCES (29s > 15.2s) and delF2 for BESS (35s < 48s), FC (35s < 48s), RFB

Impact of Energy Storage Systems in Nuclear Power Plants 129

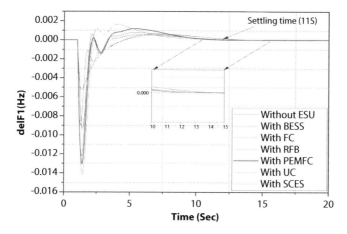

Figure 4.15a delF1 comparison of with/without ESU.

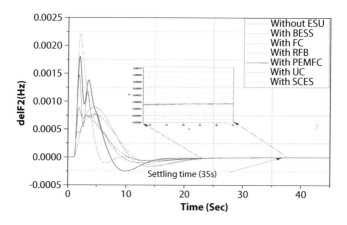

Figure 4.15b delF2 comparison of with/without ESU.

(41s < 48s), PEMFC (35s <48s), UC (35s< 48s), and SCES (43s > 48s). It is strongly identified that PEM-FC performed well over other ESUs. A bar chart comparison of the del F1 and del F2 settling time is given in Figure 4.15d, it conformed to the supremacy of the PEM-FC over other ESUs.

Bar chart in Figure 4.15d clearly demonstrates that the PEM – FC provides superior impact and support to the proposed optimization technique tuned controller for the LFC.

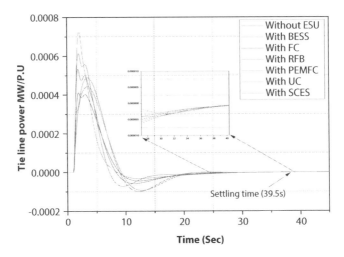

Figure 4.15c Assessment of tie line power deviation with/without ESU.

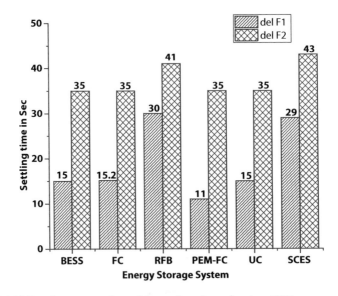

Figure 4.15d Bar chart comparison of the settling time of various ESUs.

4.7 Conclusion

In this chapter, the impact of ESU for implementing the LFM of interconnected nuclear power plants is studied. To implement the LFC in the proposed power system (an interconnected nuclear power plant), ACO tuned PID controller utilized as secondary controller. The energy storage units are utilized as small power sources, and the impact is analyzed in detail. Among various energy storage units (BESS, FC, RFB, PEM-FC, UC, SCES), the proton exchange membrane (PEM) based FC is more supported by the proposed optimization technique-based controller. While incorporating the PEM-FC in the proposed power system the secondary controller settles the oscillation in the frequency (F1 = (PEM-FC)11s < (BESS)15s < (FC)15.2s < (UC)15s < (SCES)29s < (RFB)30s) and (F2 = (PEM-FC = RFB = UC = BESS)35s < (RFB) 41s < (SCES) 43s) and also provide minimal peak values.

Challenges of energy storage units in power system

- The energy storage technology is one of the encouraging technologies in the application of renewable energy-based energy generation, distributed generation, microgrid, and smart grid. The energy storage unit's implementation has two major challenges; one is technology based and the other is economically based.
- The implementation of the energy storage units has some limitations like power capacity, the lifespan of unit, economic, and security.
- In the economic aspect, high cost and an unhealthy market are the major factors when implementing the energy storage units in the power system.

Future scope of energy storage units in the power system

- The future of electricity mostly depends on renewable energy sources. Most renewable energy sources produce the non-linear power output. The non-linear power may create fluctuation in the power supply.
- To control the fluctuations, energy storage units are the most needed.
- In the future, the power grid may be replaced by a small decentralized or microgrid power system. In discussions about the microgrid, the energy storage units play a major role.

Appendix

$T_{BESS} = 0.1s$, $T_{FC} = 0.26s$, $T_{i/c} = 0.004s$, $T_{inv} = 0.04s$, $T_{di} = 1s$, $T_{ri} = 0.78s$, $T_1 = 0.28s$, $T_2 = 0.025s$, $T_3 = 0.04s$, $T_4 = 0.39s$, $K_{sc} = 0.5$ P.U, $T_{sc} = 0.046$, $K_{UC} = 0.7$ P.U, $T_{UC} = 0.9s$.

References

1. Sahu, P.C., Mishra, S., Prusty, R.C. and Panda, S., 2018. Improved-salp swarm optimized type-II fuzzy controller in load frequency control of multi area islanded AC microgrid. *Sustainable Energy, Grids and Networks*, 16, pp. 380-392.
2. Akram, U., Nadarajah, M., Shah, R. and Milano, F., 2020. A review on rapid responsive energy storage technologies for frequency regulation in modern power systems. *Renewable and Sustainable Energy Reviews*, 120, p. 109626.
3. Yildirim, B., 2021. Advanced controller design based on gain and phase margin for microgrid containing PV/WTG/Fuel cell/Electrolyzer/BESS. *International Journal of Hydrogen Energy*, 46(30), pp. 16481-16493.
4. Sharma, M., Bansal, R.K., Prakash, S. and Asefi, S., 2021. MVO algorithm based LFC design of a six-area hybrid diverse power system integrating IPFC and RFB. *IETE Journal of Research*, 67(3), pp. 394-407.
5. Mudi, J., Shiva, C.K., Vedik, B. and Mukherjee, V., 2021. Frequency stabilization of solar thermal-photovoltaic hybrid renewable power generation using energy storage devices. *Iranian Journal of Science and Technology, Transactions of Electrical Engineering*, 45(2), pp. 597-617.
6. Dhanasekaran, B., Siddhan, S. and Kaliannan, J., 2020. Ant colony optimization technique tuned controller for frequency regulation of single area nuclear power generating system. *Microprocessors and Microsystems*, 73, p. 102953.
7. Shankar, R., Chatterjee, K. and Bhushan, R., 2016. Impact of energy storage system on load frequency control for diverse sources of interconnected power system in deregulated power environment. *International Journal of Electrical Power & Energy Systems*, 79, pp. 11-26.
8. Xu, D., Liu, J., Yan, X.G. and Yan, W., 2017. A novel adaptive neural network constrained control for a multi-area interconnected power system with hybrid energy storage. *IEEE Transactions on Industrial Electronics*, 65(8), pp. 6625-6634.
9. Boopathi, D., Jagatheesan, K., Anand, B., Kumarakrishnan, V. and Samanta, S., 2022. Effect of Sustainable Energy Sources for Load Frequency Control (LFC) of Single-Area Wind Power Systems. In *Industrial Transformation* (pp. 87-98). CRC Press.

10. Sadoudi, S., Boudour, M. and Kouba, N.E.Y., 2021. Multi-microgrid intelligent load shedding for optimal power management and coordinated control with energy storage systems. *International Journal of Energy Research*.
11. Ali, M., Parwanti, A. and Cahyono, I., 2018, October. Comparison of LFC Optimization on Micro-hydro using PID, CES, and SMES based Firefly Algorithm. In *2018 5th International Conference on Electrical Engineering, Computer Science and Informatics (EECSI)* (pp. 204-209). IEEE.
12. Zhou, X., Dong, C., Fang, J. and Tang, Y., 2017, October. Enhancement of load frequency control by using a hybrid energy storage system. In *2017 Asian Conference on Energy, Power and Transportation Electrification (ACEPT)* (pp. 1-6). IEEE.
13. Abazari, A., Monsef, H. and Wu, B., 2019. Coordination strategies of distributed energy resources including FESS, DEG, FC and WTG in load frequency control (LFC) scheme of hybrid isolated micro-grid. *International Journal of Electrical Power & Energy Systems*, 109, pp. 535-547.
14. Khadanga, R.K. and Kumar, A., 2019. Analysis of PID controller for the load frequency control of static synchronous series compensator and capacitive energy storage source-based multi-area multi-source interconnected power system with HVDC link. *International Journal of Bio-Inspired Computation*, 13(2), pp. 131-139.
15. Arya, Y., 2019. AGC of PV-thermal and hydro-thermal power systems using CES and a new multi-stage FPIDF-(1+ PI) controller. *Renewable Energy*, 134, pp. 796-806.
16. Ali, M., Djalal, M.R., Fakhrurozi, M. and Ajiatmo, D., 2018, October. Optimal Design Capacitive Energy Storage (CES) for Load Frequency Control in Micro Hydro Power Plant Using Flower Pollination Algorithm. In *2018 Electrical Power, Electronics, Communications, Controls and Informatics Seminar (EECCIS)* (pp. 21-26). IEEE.
17. Sharma, R.K. and Mishra, S., 2017. Dynamic power management and control of a PV PEM fuel-cell-based standalone ac/dc microgrid using hybrid energy storage. *IEEE Transactions on Industry Applications*, 54(1), pp. 526-538.
18. Jagatheesan, K., Anand, B., Dey, N. and Ebrahim, M.A., 2016. Design of proportional-integral-derivative controller using stochastic particle swarm optimization technique for single-area AGC including SMES and RFB units. In *Proceedings of the Second International Conference on Computer and Communication Technologies* (pp. 299-309). Springer, New Delhi.
19. Jagatheesan, K., Anand, B., Dey, N. and Ashour, A.S., 2018. Effect of SMES unit in AGC of an interconnected multi-area thermal power system with ACO-tuned PID controller. In *Advancements in Applied Metaheuristic Computing* (pp. 164-184). IGI Global.
20. Boopathi, D., Saravanan, S., Jagatheesan, K. and Anand, B., 2021. Performance Estimation of Frequency Regulation for a Micro-Grid Power System Using PSO-PID Controller. *International Journal of Applied Evolutionary Computation (IJAEC)*, 12(2), pp. 36-49.

21. Kumarakrishnan, V., Vijayakumar, G., Boopathi, D., Jagatheesan, K., Saravanan, S. and Anand, B., 2020. Optimized PSO Technique Based PID Controller for Load Frequency Control of Single Area Power System. *Solid State Technology*, 63(5), pp. 7979-7990.
22. Dhundhara, S. and Verma, Y.P., 2018. Capacitive energy storage with optimized controller for frequency regulation in realistic multisource deregulated power system. *Energy*, 147, pp. 1108-1128.
23. Jin, T.H., Chung, M., Shin, K.Y., Park, H. and Lim, G.P., 2016. Real-time dynamic simulation of korean power grid for frequency regulation control by MW battery energy storage system. *Journal of Sustainable Development of Energy, Water and Environment Systems*, 4(4), pp. 392-407.
24. Kouba, N.E.Y., Menaa, M., Hasni, M. and Boudour, M., 2016. A novel optimal frequency control strategy for an isolated wind–diesel hybrid system with energy storage devices. *Wind Engineering*, 40(6), pp. 497-517.
25. Kumar, R. and Sharma, V.K., 2020. Whale optimization controller for load frequency control of a two-area multi-source deregulated power system. *International Journal of Fuzzy Systems*, 22(1), pp. 122-137.

5

Detailed Comparative Analysis and Performance of Fuel Cells

Tejinder Singh Saggu[1]* and Arvind Dhingra[2]

[1]*Electrical Engineering Department, Punjab Engineering College, Chandigarh, India*
[2]*Electrical Engineering Department, Guru Nanak Dev Engineering College, Ludhiana, India*

Abstract

The depleting conventional sources of energy have put strains on the availability of power in the energy-savvy world. Newer sustainable sources of energy need to be harnessed to meet the ever-growing demand for energy. Fuel cells are one such source, which is non-polluting and sustainable. Fuel cells provide a technological idea for a potentially wide variety of future applications in energy storage applications, which could include on-site electric power for our houses and commercial buildings. They are used in different fields of applications like industrial, residential, commercial, and transportation. Their efficiency is very high as compared to ordinary combustion engines and standard batteries with almost zero emissions. During their operation, there is a complete absence of smog because of the lack of various air pollutants. In this chapter, a detailed description and process of various fuel cells along with their field of applications is presented. This comparison is based upon different parameters as per their field of application. This information is essential for the researchers at the beginning level so that they can compare various fuel cells and choose their occupation of the application accordingly.

Keywords: Fuel cells, hydrogen, electrolyte, methanol, alkaline, sustainability

5.1 Introduction

It was Sir Humphrey Davy who gave conceptually the first design of a fuel cell in 1802. Although he could see the potential of it, he was not able to

---

Corresponding author: saggutejinder@gmail.com

Sandeep Dhundhara, Yajvender Pal Verma, and Ashwani Kumar (eds.) *Energy Storage Technologies in Grid Modernization*, (135–158) © 2023 Scrivener Publishing LLC

designate the procedure of a carbon fuel cell which would be operational at average room temperature. The electrolyte used in this case was nitric acid. After some time, in 1839, Sir William Grove successfully operated a hydrogen-oxygen fuel cell. In this fuel cell, the chemical reaction was occurring between the oxygen and Hydrogen, and the by-products were water and electricity. This cell re-emphasized the principle of regeneration. Developments continued, and today we have an extensive number of fuel cells that are providing energy for running cars also. Fuel cells have a wide range of applications in several fields. They are beneficial in locations where we do not want pollution because they are a clean source of energy. They are used in military submarines and space planes, among other things. They are in high demand since they are not dependent on fossil fuels, which are rapidly decreasing. Fuel cell technology is designated as the most efficient and clean technology for power generation purposes in a number of applications such as automobile and aircraft systems. This is being done in order to save the environment from a large number of fossil fuels burning. In automobile applications, PEM (polymer electrolyte membrane) fuel cells provide very favorable technology [1, 2]. Another combination of PEM fuel cells along with lithium-ion batteries is also possible in hybrid vehicle technology [3]. The numerous advantages of fuel cells have made them suitable to be used in the field of medical applications as well [4].

The energy storage element in the hybrid system consists mainly of batteries or an ultra-capacitor. The response time of an ultra-capacitor is higher compared to the battery [5]. Some researchers have used several optimization techniques like Genetic algorithms, which will make a hybrid system more efficient and increases its life span as well [6, 7]. To make the system more durable, some researchers have used various modelling tools according to the various components involved in the design process and different operating conditions [8, 9]. Another possibility is to remove organic pollutants like carbohydrates, amino acids and fatty acids from the wastewaters and convert the amount of chemical energy stored in them. For this purpose, a photocatalytic fuel cell can be used [10]. Fuel cells are an efficient technique to supply continuous power to microgrid applications. Fuel cell microgrids can be run independently from the utility power grid using this technology [11, 12].

5.2 Classification of Fuel Cells

The fuel cell works on the principle of converting chemical energy to electrical energy. Usually, hydrogen and oxygen gases are supplied as reactants and electricity, and water are the output available. Hydrogen is used because of the following benefits:

- Hydrogen is the lightest of all gases (Atomic Weight = 1)
- Immense energy content (CV = 143MJ/kg) compared to petrol (CV = 43 MJ/kg), diesel (CV = 45 MJ/kg) and CNG (CV = 52 MJ/kg)
- Low energy density (ρ = 0.019 kg/m^3) at STP
- Availability in abundance
- ZERO Emissions

There are several types of fuel cells that differ in their operating temperature, electrolyte, and the fuel used. Here are the most common types of fuel cells:

- Polymer Electrolyte Membrane Fuel Cells (PEMFC): PEMFCs are the most common type of fuel cell, and they use a polymer membrane as an electrolyte. PEMFCs operate at low temperatures, typically between 60-80°C. They are widely used in transportation applications, such as cars, buses, and forklifts, as well as in portable power applications, such as laptops and mobile phones.
- Solid Oxide Fuel Cells (SOFC): SOFCs use a solid ceramic electrolyte and operate at high temperatures, typically between 800-1000°C. They are highly efficient and can use a variety of fuels, such as natural gas, biogas, and hydrogen. SOFCs are commonly used in large-scale power generation applications, such as in combined heat and power (CHP) systems for buildings and in utility-scale power plants.
- Alkaline Fuel Cells (AFC): AFCs use an alkaline electrolyte and operate at high temperatures, typically between 90-100°C. They were first developed for use in space missions, but are now being researched for use in other applications, such as stationary power generation and submarines.
- Phosphoric Acid Fuel Cells (PAFC): PAFCs use phosphoric acid as the electrolyte and operate at higher temperatures than PEMFCs, typically between 150-200°C. They are commonly used in stationary power generation applications, such as in hospitals and office buildings.

- Molten Carbonate Fuel Cells (MCFC): MCFCs use a molten carbonate electrolyte and operate at high temperatures, typically between 600-700°C. They are commonly used in large-scale power generation applications, such as in factories and utility-scale power plants.

Each type of fuel cell has its own advantages and disadvantages, and the choice of fuel cell technology depends on the specific application and requirements. The types of fuel cells and their functioning with applications are discussed in the following sections.

5.2.1 Based on Fuel-Oxidizer Electrolyte

5.2.1.1 Direct Fuel Cell

In this type of fuel cell, fuel and an oxidizing agent are used for the conversion from chemical energy to electrical energy. Fuel cells provide a good solution for processes that demand very high efficiency with low pollution. Some key advantages regarding FCs are:

i. Efficiency is more elevated than conventional power plants (30-40%).
ii. Fuel efficiency can be improved to 90% if waste heat is utilised in the system.
iii. Low pollution level.
iv. Low noise level.
v. Fuel cell requires low maintenance.
vi. Cogeneration of heat can increase the efficiency.

Fuel cells (FCs) are categorised in various ways depending on various parameters as discussed above under different operating conditions and temperatures [13]. Direct Fuel Cells (DFCs) are a very economical and emerging type in nature; they have very high energy density and instant recharge capability and they are easily transportable. They have many advantages over Hydrogen Fed Polymer Electrolyte Membrane Fuel Cell (PEMFC). In PEMFC, the fuel used is mainly Hydrogen and it is primarily used in transportation applications [14].

It can be noted that Hydrogen is difficult to transport and can be highly flammable. Also, due to storage problems, DFCs appear to be an excellent alternative to various other fuel cells. Also, because of its easy handling, it is most suitable in the application demanding around 1-3 kW of electrical energy. Furthermore, the devices which operate lithium-ion batteries are difficult to move because of a lack of power supply. Figure 5.1 below shows the functional diagram of DFCs. In DFCs, the oxidation process occurs towards the anode side and liquid fuel is supplied to this side as well. The reduction process happens towards the cathode side, and oxygen gas or air is provided to this side. Hence, electrons pass through an external circuit because charged ions move via the electrolyte. Therefore, the waste product is water towards the cathode, and due to incomplete fuel oxidation, there is some by-product towards the anode side [15]. Due to the high power output of DMFC fuel cells, they are mainly used in various portable equipment [16]. Following are the types of Direct Fuel Cells:

   i. Methanol-based fuel cells (DMFCs)
   ii. Ethanol-based fuel cells (DEFCs)
   iii. Ethylene glycol–based fuel cells (DEGFCs)
   iv. Glycerol-based fuel cells (DGFCs)
   v. Formic acid–based fuel cells (DFAFCs)
   vi. Dimethyl ether–based fuel cells (DDEFCs)
   vii. Hydrazine-based acid fuel cells (DHFCs)

Classification based on Temperature

As discussed earlier, the fuel cells are also categorised as per the various operating temperature conditions. For example, SOFC having solid electrolytes basically operates at high temperatures. The detailed review of working temperature, electrolyte, oxidant used and various issues related to fuel cells is tabulated in Table 5.1.

Although a lot of research is going on in the development of an AFC, it is still not favourable to be used as compared to a PEMFC. This is mainly due to some performance instability issues and improper water management within the fuel cell [17]. To make the fuel cell more durable and to improve its performance, fuel recirculation can also be considered. This can be done in two ways: Active and passive fuel recirculation [18].

5.2.1.2 Regenerative FC

A fuel cell that can act in reverse mode is also called a Regenerative Fuel Cell (RFC). It can work in two modes:

1) Power Generation Mode
2) Reverse Mode

In the first mode, it produces electric energy from the fuel cell, while in the second mode, it generates fuel from electricity [19–21].

Table 5.1 Categorisation of direct fuel cells.

Parameter	Solid oxide fuel cell (SOFC)	Molten carbonate fuel cell (MCFC)	Phosphoric acid fuel cell (PAFC)	Proton exchange membrane fuel cell (PEMFC)	Alkaline fuel cell (AFC)
Temperature (°C)	800-1000	600-700	160-220	80-110	60-90
Oxidant	Oxygen or air	Oxygen, CO_2 or air	Oxygen or air	Oxygen or air	Oxygen or air
Fuel Used	H_2/CO/CH4	H_2/CO	H_2 reformate	H_2 reformate	H_2
Electrolyte	ZrO_2 with $Y2O_3$	$LiCO_3$-K_2CO_3	H_3PO_4	Polymer membrane	NaOH/ KOH
Key issues	Ceramic cells	CO_2 recycling necessary	CO sensitivity	Fuel Moisture	CO_2 troubles
Efficiency (%)	60-65	55-65	50-55	55-60	55-60
Applications	Distributed Generation, Electrical utility, Providing auxiliary power	Distributed Generation, Electrical utility	Distributed Generation	Transportation, Power Backup, Distributed Generation	Transportation, Power Backup, Military, Space

(Continued)

Table 5.1 Categorisation of direct fuel cells. (*Continued*)

Parameter	Solid oxide fuel cell (SOFC)	Molten carbonate fuel cell (MCFC)	Phosphoric acid fuel cell (PAFC)	Proton exchange membrane fuel cell (PEMFC)	Alkaline fuel cell (AFC)
Advantages	Fuel flexibility, High efficiency, Suitable for Cogeneration, Solid electrolyte	Fuel flexibility, High efficiency, Suitable for Cogeneration	Suitable for Cogeneration, Enhanced fuel impurities tolerance	Faster start-up, Low temperature, Less electrolytic management problems	Faster start-up, Low temperature, Lower cost components
Challenges	Long start-up time, Limited number of shut downs, High temperature corrosion of components	Long start-up time, High temperature corrosion of components, Low power density	Expensive catalysts, Long start-up time, Sulphur sensitivity	Sensitive to fuel impurities, Expensive catalysts	Electrolyte management and conductivity problems, Sensitive to CO_2 in fuel and air

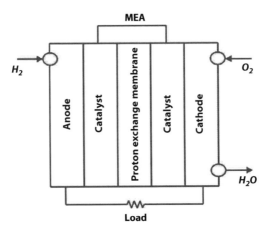

Figure 5.1 Working of regenerative FC in fuel cell mode.

Mode 1: Power Generation Mode (Fuel Cell Mode)

In this mode, as shown in Figure 5.1, energy hydrogen fuel converts electricity using oxygen. Hydrogen, and oxygen is supplied on anode and cathode, respectively. The chemical reaction in this mode is as follows:

$$H_2 + \frac{1}{2}O_2 \rightarrow H_2O + Electrical\ Energy + Heat \tag{5.1}$$

Mode 2: Reverse Mode

Energy of DC source and environmental heat is given to RFC in this mode. This allows RFC to decompose water into oxygen and hydrogen fuel. Figure 5.2 shows the basic diagram of this mode. In both modes, two

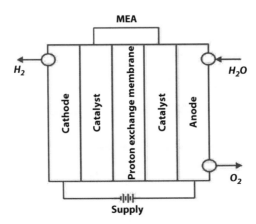

Figure 5.2 Working of RFC in reverse mode.

catalysts and Proton Exchange Membrane forms Membrane Exchange Assembly (MEA). After that, Catalysts and MEA start the reaction. The MEA and Catalysts are sandwiched between plates. These plates are called cathode and anode field plates. It has many functions, including distribution of fuel, heat management facilitation, etc.

$$H_2O + Electrical\ Power + Heat \rightarrow H_2 + \frac{1}{2}O_2 \qquad (5.2)$$

5.2.1.3 Indirect Fuel Cells

In the case of direct cells, the hydrocarbon fuel injects straightway into the fuel cell. However, in the case of indirect fuel cells, which are also known as reformed fuel cells, the hydrocarbon fuel is reformed initially to remove the Hydrogen and then it is injected into the fuel cell. The place where the Hydrogen is removed from the hydrocarbons is known as a reformer, which is also known as a separate fuel processing station. On the anode side, the Hydrogen is first broken down into an electron and a proton and further reformed on the cathode side for the production of water, as shown in Figure 5.3 below.

Thus, we can also term indirect fuel cells as the simple hydrogen fuel cells that receive Hydrogen by reforming the hydrocarbons directly on the plant site. Methanol and ethanol are commonly preferred in this process for the production of Hydrogen. Both of these provide better energy having suitable Hydrogen carbon ratio for the same amount of CO_2 released.

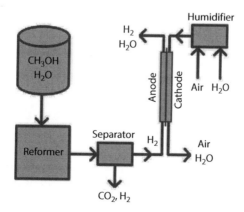

Figure 5.3 Indirect fuel cell operation (Source: iconspng.com).

Advantages:

1. There is no cost expenditure for transportation of Hydrogen and thus no risks involved either.
2. There are no problems due to catalyst poisoning like indirect fuel cells.
3. There is no particular requirement for a water supply and management system as the water is not being supplied with hydrocarbon in the fuel cell.
4. Unlike the direct fuel cells, the reforming process is controlled separately in these indirect FCs, hence these types of cells are able to operate at lesser temperatures.

Limitations:

1. Indirect fuel cells are challenging to maintain as they are more complex.
2. They are not so reliable as direct fuel cells.
3. The heat generated by the reformer can be used for some additional applications.

5.2.2 Based on the State of Aggregation of Reactants

5.2.2.1 Solid Fuel Cells

When some complex ceramic compounds like zirconium oxide or calcium oxide are used as electrolytes, the cell is known as a SOFC, as shown in Figure 5.4 below. Such cells operate at around 1000 °C temperature, giving an output of 100 kW with an efficiency of 60-70%. There is no need for a reformer to extract Hydrogen from the fuel at such a high temperature. Thus, electricity can also be produced by recycling the waste heat. Leakage of electrolytes is not possible in these cells like liquid and gaseous fuel cells, but a very high temperature limits the applications of these cells.

In SOFC, the electrolyte used in the cell is of solid oxide material, which conducts negative oxygen ions to the anode from the cathode. On the anode side, the oxidation of carbon monoxide or Hydrogen takes place with the oxygen. The SOFC operates at a very high temperature; hence there is no need for any expensive platinum catalyst. So, instead of a membrane or liquid, zirconium oxide is used as a solid-state catalyst mixed with yttrium oxide. The reaction kinetics are also improved by the application of high temperature having no need for a metal catalyst.

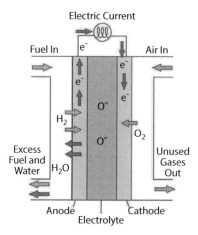

Figure 5.4 SOFC (Source: Fuel cell Store).

SOFCs are classified into mainly three geometries; micro-tubular, planar and coplanar. The air flows inside the solid oxide tube sealed on one side and on the other side, and the fuel flows outside in the tubular design. The cell consists of different layers, and the cathode is represented by the tube itself. Whereas the planar design consists of different flat stacks in which the Hydrogen and air flow through different channels representing cathode and anode. This type of SOFC is mainly used in small and large power plants and cogeneration plants for the production of heat and electricity.

5.2.2.2 Gaseous Fuel Cells

A gaseous FC makes use of a Proton Exchange Membrane (PEM) unit, which uses mainly two types of gases as a fuel; Hydrogen and Oxygen. The final reaction using these gases involves heat, water and electricity as by-products. This technique is very much more straightforward in nature than the other forms of power plants like thermal power plants, nuclear power plants, diesel engines, etc. All these processes cause a lot of pollution and produce detrimental by-products. So, in a gaseous fuel cell, the electrolysis process is usually preferred for supplying Hydrogen; the other gas oxygen is readily available in the atmosphere.

A PEM unit consists of mainly four elements: Anode, cathode, electrolyte and catalyst, as shown in Figure 5.5. The negative terminal of the cell is known as the anode. Its primary function is to collect the electrons from the hydrogen molecules which are used in an external circuit. The hydrogen

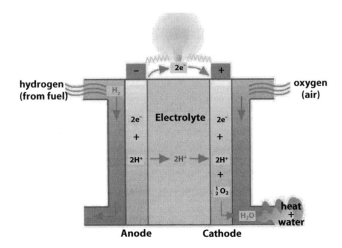

Figure 5.5 A PEM fuel cell (Source: World Fuel Cell Council).

gas is equally dispersed in the catalyst surface using different channels imprinted on this unit. On the other hand, the positive terminal of the cell is called the cathode. It has separate channels branded to dispense the oxygen molecules over the surface of the catalyst. Electrons from the outside surface are conducted back into the catalyst, where the recombination of Hydrogen and oxygen molecules takes place to produce water.

The PEM unit is basically acting as an electrolyte; whose main function is to conduct positively charged ions. The electrons are basically blocked by the membrane. Therefore, for it to remain in a stable state and function properly, a membrane should always be hydrated. However, the reaction of Hydrogen and oxygen takes place using a catalyst consisting of platinum nanoparticles coating facing the PEM unit. The surface of a catalyst is usually rough and absorbent. This is being done in order to expose the maximum surface area of platinum with oxygen and Hydrogen. The hydrogen PEM fuel cells are mainly used in EV applications. A lot of research is going on for commercializing the manufacturing of PEM fuel cell stacks by examining the various types of barriers in this process [22–24]. The various reactions taking place in a PEM unit are as shown below:

$$\textbf{Anode: } 2H_2 = 4H+ + 4e- \tag{5.3}$$

$$\textbf{Cathode: } 4H+ + 4e- + O_2 = 2H_2O \tag{5.4}$$

5.2.2.3 Liquid Fuel Cells

In fuel cells, there are plenty of good options available to use the liquid as a fuel. The compressed Hydrogen, Methanol, Ethanol & Ammonia are the primary fuels which are used as a liquid in fuel cells. However, a fuel cell is not so efficient when we use pure Hydrogen as a fuel. The hydrogen infrastructure cost is also very high. A number of sources like hydrocarbons, natural gas, sunlight etc., may be used to produce Hydrogen. This is the primary fuel which has been used to launch space shuttles by NASA since 1970. Electricity, heat and water are produced by mixing Hydrogen with oxygen. A lot of research is going on in this area to utilize pure Hydrogen as a fuel from non-conventional energy sources. Hydrogen is also finding various uses in electric vehicles, producing no pollution at all [25]. The main limitation in this type of cell is the high cost of catalyst as well as its loading. Two main types of liquid fuels using Methanol and ethanol are mainly used for commercial applications [26]. Liquid FCs can also be produced through solar power, and the whole process of this conversion is described in this research article [27].

However, Methanol, Ethanol & Ammonia can be used as such in the fuel cells. One such example is shown in Figure 5.6. They can also be processed outside the fuel cell using a reformer. This is being done to enhance the life of the fuel cell catalyst.

Methanol is an alcohol-based fuel having an energy density higher than compressed Hydrogen. It is also a safer fuel as compared to Hydrogen. It is also readily available from many supply chains. Natural gas, Coal & Biomass are the main sources of producing Methanol across the world.

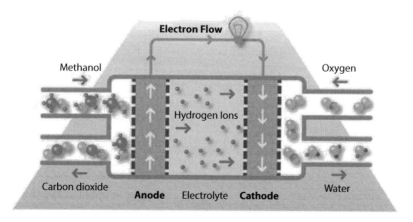

Figure 5.6 Liquid fuel cells using Methanol (Source: Fuel Cell Store).

Research shows that fuel cells powered by Methanol are highly preferred for commercial fuel cells. Ethanol as liquid fuel is also an attractive choice because of several advantages like better supply chain, high energy density, lower infrastructure cost, less storage requirements and safer as well. It can be produced from fossil fuels, feedstocks and a variety of biological sources using fermentation and distillation processes. Various plants like corn, sugarcane and switchgrass are also used for its production. Ammonia is also a good option in liquid fuel cells due to its clean combustion properties because only Hydrogen and nitrogen are left as by-products during its reformation process.

It contains 17.6% of hydrogen atoms, quite similar to the Methanol weight content during partial oxidation reformation, thereby making it an ideal carbon-free fuel [28–30]. By applying a pressure of 10 bar at 300 K using a liquid density of 600 g/L, it is liquefied. The cracking reaction of ammonia is given as:

$$NH_3 = 0.5\ N_2 + 1.5\ H_2\ (\text{at } 400\ °C) \tag{5.5}$$

This high temperature can be obtained using any external source. Ammonia as fuel has one drawback of undissipated ammonia concentration of around 50 ppm during its product gas formation, which may damage the fuel cells. This may be prevented by using an acid scrubber which can quickly ease the traces of ammonia gas from the cracker.

5.2.3 Based on Electrolyte Temperature

5.2.3.1 Proton Exchange Membrane

This cell, also known as polymer electrolyte membrane FC, was discovered in 1959. The operating temperature range for this cell is 50 to 100°C. The cell construction is shown in Figure 5.7 below. The electrocatalyst used is generally platinum, and it is mainly preferred in automobile and submarine applications. The efficiency of this cell is 50 to 60%.

On the anode side, the following chemical reaction takes place:

$$H_2 \rightarrow 2H^+ + 2e^- \tag{5.6}$$

On the cathode side, the following chemical reaction takes place:

$$\frac{1}{2}O_2 + 2H^+ + 2e^- \rightarrow H_2O \tag{5.7}$$

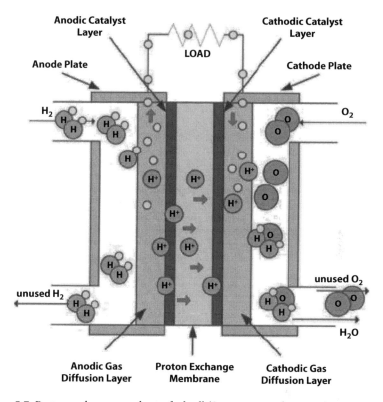

Figure 5.7 Proton exchange membrane fuel cell (Source: researchgate.net).

Thus, net electrochemical reaction is:

$$H_2 + \frac{1}{2}O_2 \rightarrow H_2O \tag{5.8}$$

The PEM FC is accepted chiefly because it takes significantly less time to start and stop, which makes it suitable to be used in remote areas. Other important parameters of a PEM fuel cell are its durability, sustainability and simplicity, which make it suitable to be used in various industrial areas. It is also widely accepted in the automotive industry and other portable equipment because of its high power density, noise-free operation, lower operating temperature and almost zero carbon emissions.

5.2.3.2 Direct Methanol

This type of cell is shown in Figure 5.8. The development of this fuel cell was done in 1960. The ceramic is being used as an electrocatalyst, and the temperature range of this cell is about 450 to 500°C. The operating efficiency is about 30 to 40%.

On the anode side, the following chemical reaction takes place:

$$CH_3OH + H_2O \rightarrow CO_2 + 6H^+ + 6e^- \tag{5.9}$$

On the cathode side, the following chemical reaction takes place:

$$\frac{3}{2}O_2 + 6H^+ + 6e^- \rightarrow 3H_2O \tag{5.10}$$

Thus, the net electrochemical reaction is:

$$CH_3OH + \frac{3}{2}O_2 \rightarrow CO_2 + 2H_2O \tag{5.11}$$

5.2.3.3 Alkaline

This type of cell is shown in Figure 5.9. The development of this fuel cell was also done in 1960. The platinum is being used as an electro catalyst and the temperature range of this cell is about 50 to 220°C. The operating

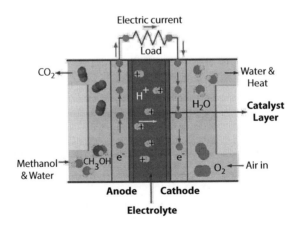

Figure 5.8 Direct methanol (Source: researchgate.net).

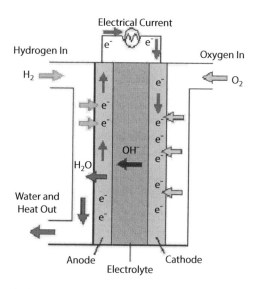

Figure 5.9 Alkaline fuel cell (Source: researchgate.net).

efficiency is about 50 to 60%. It is mainly used in transportation and space shuttle applications.

On the anode side, the following chemical reaction takes place:

$$2H_2 + 4OH^- \rightarrow 4H_2O + 4e^- \qquad (5.12)$$

On the cathode side, the following chemical reaction takes place:

$$O_2 + 2H_2O + 4e^- \rightarrow 4OH^- \qquad (5.13)$$

Thus, the net electrochemical reaction is:

$$2H_2 + O_2 \rightarrow 2H_2O \qquad (5.14)$$

5.2.3.4 Phosphoric Acid

This type of cell is shown in Figure 5.10. The development of this fuel cell was done in 1965. Generally, platinum is being used as an electrocatalyst and the temperature range of this cell is about 150 to 220°C. The power density is about 55%. It was the first FC that was used for commercial purposes.

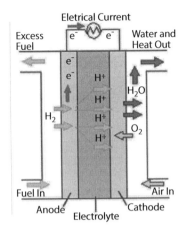

Figure 5.10 Phosphoric acid fuel cell (Source: researchgate.net).

On the anode side, the following chemical reaction takes place:

$$H_2 \rightarrow 2H^+ + 2e^- \tag{5.15}$$

On the cathode side, the following chemical reaction takes place:

$$\frac{1}{2}O_2 + 2H^+ + 2e^- \rightarrow H_2O \tag{5.16}$$

Thus, the net electrochemical reaction is:

$$H_2 + \frac{1}{2}O_2 \rightarrow H_2O \tag{5.17}$$

5.2.3.5 Molten Carbonate

This type of cell is shown in Figure 5.11. The development of this fuel cell was also done in the 1960s. The alkali carbonate is being used as an electrocatalyst, and the temperature range of this cell is about 600 to 700°C. The operating efficiency is about 55 to 65%. It is also used in commercial applications.

On the anode side, the following chemical reaction takes place:

$$H_2 + CO_3^= \rightarrow H_2O + CO_2 + 2e^- \tag{5.18}$$

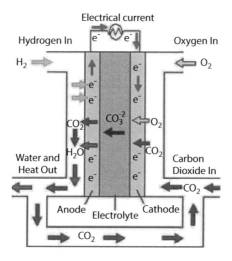

Figure 5.11 Molten carbonate fuel cell (Source: researchgate.net).

On the cathode side, the following chemical reaction takes place:

$$\frac{1}{2}O_2 + CO_2 + 2e^- \rightarrow CO_3^= \tag{5.19}$$

Thus, the net electrochemical reaction is:

$$H_2 + \frac{1}{2}O_2 + CO_2 \rightarrow H_2O + CO_2 \tag{5.20}$$

5.2.3.6 Solid Oxide

This type of cell is shown in Figure 5.12. The development of this fuel cell was also done in the 1950s. The calcium titanate is being used as an electrocatalyst, and the temperature range of this cell is about 700 to 1000°C. The operating efficiency is about 55 to 65%.

On the anode side, the following chemical reaction takes place:

$$H_2 + O^= \rightarrow H_2O + 2e^- \tag{5.21}$$

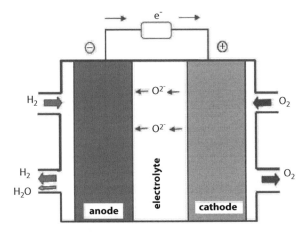

Figure 5.12 Solid Oxide Fuel cell (Source: researchgate.net).

On the cathode side, the following chemical reaction takes place:

$$\frac{1}{2}O_2 + 2e^- \rightarrow O^= \qquad (5.22)$$

Thus, the net electrochemical reaction is:

$$H_2 + \frac{1}{2}O_2 \rightarrow H_2O \qquad (5.23)$$

5.3 Cost of Different Fuel Cell Technologies

A thorough market analysis is essential when determining the best fuel cell for a specific application. This would entail determining the performance and operational needs, such as the number of operating hours, projected life duration, frequency, and so on. For efficient use of fuel cells, a proper system must be designed. This would entail determining how many cells are required, as well as which combinations of stacks are needed. The cost is the next significant consideration. The cost of a fuel cell per kW can vary widely depending on the type of fuel cell, the manufacturing process, the scale of production, and other factors. However, in general, the cost of fuel cells has been decreasing over time as technology advances and production volumes increase. As of 2021, the cost of a proton exchange membrane

(PEM) fuel cell, which is one of the most common types of fuel cells used for transportation and stationary power applications, is estimated to be in the range of 395 EUR/kW [31]. However, it is important to note that these estimates can vary widely depending on the specific application and other factors. The detailed comparative analyses of the different fuel cell technologies in term of the cost of the system is found in Refs [31, 32].

5.4 Conclusion

Nowadays, a lot of energy is required in every industry practically, putting a lot of pressure on state electricity boards and utilities to supply this demand. Fuel cells are a prominent source of energy that is both environmentally friendly and sustainable. Their efficiency is relatively high compared to traditional energy sources and conventional batteries, and they emit nearly no carbon dioxide. With the ever-increasing need for electrical energy, the impetus is on to find energy sources that are both sustainable and non-polluting. Thus, fuel cells are ideal for all these applications. This chapter attempts to describe several types of fuel cells, their operating principles, and diverse fields of use such as electric vehicles, hybrid vehicles, medical, commercial, and industrial applications. Researchers interested in working in this discipline can use this information to determine and select their field of application based on their interests.

References

1. Yang Luo, Yinghong Wu, Bo Li, Tiande Mo, Yu Li, Shien-Ping Feng, Jingkui Qu, Paul K. Chu, Development and application of fuel cells in the automobile industry. Journal of Energy Storage, 42, 103-124, 2021.
2. S. Porstmann, T. Wannemacher, W.-G. Drossel, A comprehensive comparison of state-of-the-art manufacturing methods for fuel cell bipolar plates including anticipated future industry trends. Journal of Manufacturing Processes, 60, 366-383, 2020.
3. Mohamed Nacereddine Sid, Mohamed Becherif, Abdenacer Aboubou, Amel Benmouna, Power control techniques for fuel cell hybrid electric vehicles: A comparative study. *Computers & Electrical Engineering*, 2021.
4. Qian Xu, Feihu Zhang, Li Xu, Puiki Leung, Chunzhen Yang, Huaming Li, The applications and prospect of fuel cells in medical field: A review. Renewable and Sustainable Energy Reviews, 67, 574-580, 2017.

5. K. Latha, B. Umamaheswari, K. Chaitanya, N. Rajalakshmi, K.S. Dhathathreyan, A novel reconfigurable hybrid system for fuel cell system. International Journal of Hydrogen Energy, 40, 14963-14977, 2015.
6. N. Sulaiman, M.A. Hannan, A. Mohamed, P.J. Ker, E.H. Majlan, W.R. Wan Daud, Optimization of energy management system for fuel-cell hybrid electric vehicles: Issues and recommendations. *Applied Energy*, 228, 2061-2079, 2018.
7. Xueqin Lü, Yinbo Wu, Jie Lian, Yangyang Zhang, Chao Chen, Peisong Wang, Lingzheng Meng, Energy management of hybrid electric vehicles: A review of energy optimization of fuel cell hybrid power system based on genetic algorithm. Energy Conversion and Management, 205, 112-124, 2020.
8. L. Vichard, N. Yousfi Steiner, N. Zerhouni, D. Hissel, Hybrid fuel cell system degradation modeling methods: A comprehensive review. Journal of Power Sources, 506, 230-245, 2021.
9. Julia Savioli, Graeme W. Watson, Computational modelling of solid oxide fuel cells. Current Opinion in Electrochemistry, 21, 14-21, 2020.
10. Yasser Vasseghian, Alireza Khataee, Elena-Niculina Dragoi, Masoud Moradi, Samaneh Nabavifard, Gea Oliveri Conti, Amin Mousavi Khaneghah, Pollutants degradation and power generation by photocatalytic fuel cells: A comprehensive review. Arabian Journal of Chemistry, 13, 8458-8480, 2020.
11. Subhashree Choudhury, Nikhil Khandelwal, A critical survey of fuel cells applications to microgrid integration: Configurations, issues, potential solutions and opportunities. International Transactions on Electrical Energy System, 213, 1353-1370, 2019.
12. Valverde L, F. Rosa, C. Bordons, J. Guerra, Energy Management Strategies in hydrogen Smart-Grids: A laboratory experience. International Journal of Hydrogen Energy, 41, 13715-13725, 2016.
13. B.C. Ong a, S.K. Kamarudin, S. Basri, Direct liquid fuel cells: A review. International Journal of Hydrogen Energy, 42, 322-338, 2017.
14. Yun Wang, Daniela Fernanda Ruiz Diaz, Ken S. Chen, Zhe Wang, Xavier Cordobes Adroher, Materials, technological status, and fundamentals of PEM fuel cells – A review, Materials Today, 32, 178-203, 2020.
15. PJM van Tonder, "Bipolar Plates and Flow Field Topologies for the Regenerative Fuel Cell", IEEE Africon 2011 - The Falls Resort and Conference Centre, 2011.
16. M.S. Alias, S.K. Kamarudin, A.M. Zainoodin, M.S. Masdar, Active direct methanol fuel cell: An overview, International Journal of Hydrogen Energy, 45, 19620-19641, 2020.
17. Rambabu Gutru, Zarina Turtayeva, Feina Xu, Gaël Maranzana, Brigitte Vigolo, Alexandre Desforges, A comprehensive review on water management strategies and developments in anion exchange membrane fuel cells, International Journal of Hydrogen Energy, 45, 19642-19663, 2020.

18. Junbo Hou, Min Yang, Junliang Zhang, Active and passive fuel recirculation for solid oxide and proton exchange membrane fuel cells, Renewable Energy, 155, 1355-1371, 2020.
19. W. Wiyaratn, "Review on Fuel Cell Technology for Valuable Chemicals and Energy Co-Generation", Engineering Journal, 14, 1-14, 2010.
20. F. Alcaide, P. L. Cabot, and E. J. Brillas, "Fuel cells for chemicals and energy cogeneration" Journal of Power Sources, 153, 47-60, 2006.
21. Johansson, T.B., McCormick, K., Neij, L., and Turkenburg, W. C, The Potentials of Renewable energy. Renewable Energy: A Global Review of Technologies, London, 15-47, 2006.
22. Youhyun Lee, Min Chul Lee, Young Jin Kim, Barriers and strategies of hydrogen fuel cell power generation based on expert survey in South Korea, International Journal of Hydrogen Energy, 2021.
23. Mahmoud Dhimish, Romênia G. Vieira, Ghadeer Badran, Investigating the stability and degradation of hydrogen PEM fuel cell, International Journal of Hydrogen Energy, 46, 37017-37028, 2021.
24. A. Kampker, P. Ayvaz, C. Schön, J. Karstedt, R. Förstmann, F. Welker, Challenges towards large-scale fuel cell production: Results of an expert assessment study, International Journal of Hydrogen Energy, 45, 29288-29296, 2020.
25. Junye Wang, Hualin Wang, Yi Fan, Techno-Economic Challenges of Fuel Cell Commercialization, Engineering, 4, 352-360, 2018.
26. B.C. Ong, S.K. Kamarudin, S. Basri, Direct liquid fuel cells: A review, International Journal of Hydrogen Energy, 42, 10142-10157, 2017.
27. Shunichi Fukuzumi, Production of Liquid Solar Fuels and Their Use in Fuel Cells, Joule, 1, 689-738, 2017.
28. Yuqi Guo, Zhefei Pan, Liang An, Carbon-free sustainable energy technology: Direct ammonia fuel cells, Journal of Power Sources, 476, 223-245, 2020.
29. Ahmed Afif, Nikdalila Radenahmad, Quentin Cheok, Shahriar Shams, Jung H. Kim, Abul K. Azad, Ammonia-fed fuel cells: a comprehensive review, Renewable and Sustainable Energy Reviews, 60, 822-835, 2016.
30. Osamah Siddiqui, Ibrahim Dincer, A review and comparative assessment of direct ammonia fuel cells, Thermal Science and Engineering Progress, 5, 568-578, 2018.
31. A. Kampker, H. Heimes, M. Kehrer, S. Hagedorn, P. Reims, and O. Kaul, Fuel cell system production cost modeling and analysis. *Energy Reports*, vol. 9, pp. 248–255, 2023.
32. Battelle Memorial Institute, Manufacturing Cost Analysis of 100 and 250 kW Fuel Cell Systems for Primary Power and Combined Heat and Power Applications. *U.S. Dep. Energy*, no. January, pp. 1–260, 2017.

6

Machine Learning–Based SoC Estimation: A Recent Advancement in Battery Energy Storage System

Prerana Mohapatra*, Venkata Ramana Naik N. and Anup Kumar Panda

Dept. of Electrical Engineering, National Institute of Technology, Rourkela, Odisha, India

Abstract

An energy management system has become indispensable for a microgrid to manage different distributed energy resources in order to have a modern grid-connected system. To compensate for the intermittent nature of renewables and to ensure continuity in supply to the load, energy storage systems (ESS) especially battery energy storage (BES) have emerged for grid applications. The repeated charging/discharging cycles of the battery adversely affect its operational life which decreases the overall system reliability in the long term. This chapter concentrates on the management of the BES by estimating its state of charge (SoC). SoC estimation is an imperative metric to accurately estimate the available battery capacity. Recently, machine learning (ML) based estimation techniques have gained much attention as they can solve nonlinear modeling problems and their state estimation with great accuracy. In this study, ML techniques like support vector regression (SVR), and extreme learning machine (ELM) are investigated. To further improve the performance of ELM, regression analysis is also performed by using a penalty factor that reduces the error coefficient estimate toward zero. In this chapter, an extension of linear regression, i.e., ridged regression, is proposed that uses multiple regression data to minimize the prediction error. Hence, a comparative investigation of SVR and ELM with ridged regression is presented here.

---

*Corresponding author: 519ee1015@nitrkl.ac.in

Sandeep Dhundhara, Yajvender Pal Verma, and Ashwani Kumar (eds.) *Energy Storage Technologies in Grid Modernization*, (159–180) © 2023 Scrivener Publishing LLC

Keywords: Battery, energy management system, energy storage system, extreme learning machine, machine learning, state of charge, support vector regression

6.1 Introduction

The prevailing energy market scenario indicates the global proliferation in energy demand which is mostly fulfilled by fuel-based power plants. This raises serious ecological disturbances like the greenhouse effect and global warming resulting from high carbon emissions [1]. On the other hand, renewable energies are clean and adequately available. As a result, integrating renewable energy sources (RES) into the existing system has become critical. RES like solar PV and wind are mostly used for grid integration. However, due to the irregular nature of renewables, an energy storage system (ESS) becomes imperative in providing energy in the absence of RES [2]. It also helps in peak shaving during high demand like in the evening; black-out capability or enables islanding operation thereby refining the robustness of the existing system. This aids in integrating additional renewable energy sources (RES) into the energy market.

In terms of power capacity and operational time, extensive research and development have been carried out on ESS technology, e.g., compressed air energy storage, supercapacitors, electrochemical energy storage (otherwise referred to as battery energy storage (BES)), flywheels, and superconducting magnetic energy storage (SMES) [3]. These storage technologies are capable of frequent and quick charging-discharging cycles while obtaining high efficiency, and hence are applicable for power grid applications and transportation systems. Several prevalent BES devices are explained in [4] highlighting the characteristics, merits, demerits, and applications of the topologies. The selection of a particular battery type is based on the initial, installation, and maintenance cost, lifecycle, and efficiency [5]. Lithium-ion batteries are at the forefront in terms of dynamics, energy density, and cost competitiveness. They have higher cell voltage, longer lifetime, higher energy density, and remarkable life cycle, and therefore replace lead-acid batteries for several applications [6].

To improve the reliability and flexibility of the existing utility grid, the concept of a microgrid is introduced. In general, a microgrid is a scaled-down version of the utility grid that generates, manages, and controls different energy resources for a small locality. This increases safety from the cyber-threats and resistibility from any grid faults or black-out conditions as it can operate separately on its own. A typical microgrid consists of different distributed energy resources (DERs) like distributed generations

(DGs), ESS, loads, and a centralized control as seen in Figure 6.1. The chapter focuses on the BES amongst all ESS technologies.

The microgrid control mechanism includes an automated energy management system to control and coordinate the operations of different DERs [7]. This would maximize the penetration of RESs, minimize the energy losses, operational costs, fuel consumption, and carbon emission. Also, this establishes an interaction with the existing utility grid for the import/export of power and schedules the operations of DERs for proper load sharing. To achieve these objectives, an energy management system considers a few inputs from the DERs and load as listed in Figure 6.2. This chapter focuses on the energy management of the BES for appropriate power flow in the microgrid by estimating SoC accurately [8].

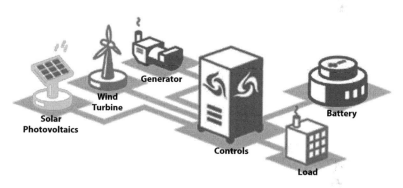

Figure 6.1 Centralized control of a microgrid.

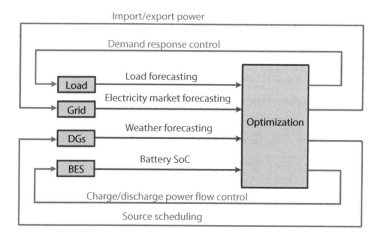

Figure 6.2 Layout of an energy management system.

For having a good life cycle, it is crucial to know the energy availability in a battery concerning the initial energy at the fully charged state. This gives the operator an estimate of the time a battery will last before needing to be recharged. A parameter called state of charge (SoC) is defined to evaluate the energy obtainability of a battery and is expressed with respect to the rated capacity as in (6.1).

$$SoC = \frac{Q_{obtainable}}{Q_{rated}} \qquad (6.1)$$

where $Q_{obtainable}$ is the maximum available energy and Q_{rated} denotes the rated battery energy capacity.

Overcharging and over-discharging scenarios may also be avoided through SoC estimation, which can improve the battery's life cycle. Therefore, an accurate SoC estimation is highly critical for user convenience and battery longevity. However, due to the unpredictability of the battery's intrinsic chemical properties, estimating SoC is challenging. This chapter concentrates on the SoC estimation of lithium-ion batteries. Numerous strategies for estimating battery SoC have been described, including the coulomb counting method, look-up table approach, model-based approach, data-driven, and hybrid approaches [9]. One of the simplest methods is the coulomb counting method. It calculates the SoC by integrating the discharging current over the cycle. However, this approach can be reliable only if the initial SoC is known and the current is precisely calibrated. The model-based approach is quite reliable and accurate but requires extensive domain knowledge, which increases the time taken for its development. The electrochemical model has high precision due to the involvement of internal chemical reactions but this also owes to high error due to chemical complexity. The SoC of a lithium-ion battery is estimated using a variety of model topologies [10]. An electrical equivalent model based on open circuit voltage is presented [11]. Alternatively, the data-driven approach needs limited prior knowledge of the internal chemical characteristics, complicated chemical reactions, and the battery's model parameters. Hence, it requires less time to model a complex system. Fundamentally, a data-driven technique can work effectively for a huge available data [12]. Artificial neural networks, fuzzy logic controllers, and machine learning (ML) techniques are a few examples of data-driven techniques that depend on the historical experimental data for training and estimating SoC. Hybrid approaches are followed by combining the advantages of any two methods. Generally, the simplicity of the coulomb counting method is applied along with the accurate model-based techniques or data-driven techniques. A similar approach is executed in [13] combining the fuzzy logic controller to estimate the SoC. Another hybrid technique is proposed in [14] which uses the k-nearest neighbor algorithm along with the Gaussian filter for lithium-ion batteries. An ELM-based SoC estimate method is presented in [15] in which the parameters are optimized using a gravitational search algorithm.

Recently, machine learning–based estimation techniques have gained much attention as they can efficiently solve nonlinear problems with great accuracy. It is realized that the regression analysis in the machine learning technique is used majorly for prediction analysis, forecasting, and estimating future data based on various dependent and independent data. Further to improve the performance, an extension of linear regression, i.e., ridged regression is proposed that introduces a small bias called penalty factor for ridge regression to minimize the prediction error. In this chapter, the ML techniques like support vector machine (SVM) and extreme learning machine (ELM) are investigated for SoC estimation. SVM applies the minimization principle to deal with a small number of samples [16, 17]. However, numerous factors, such as error control parameters and penalty coefficient, must be fixed, which takes time and makes practical implementations challenging. Conversely, ELM has a simple structure, and fast learning capability, and it is easy to regulate the parameters without getting trapped into a local minimum [18]. Hence, superiority in performance of ELM with ridged regression is evaluated here.

6.2 SoC Estimation Techniques

The classification of techniques employed for the SoC assessment is represented in Figure 6.3. A brief review of the below-mentioned methods is presented highlighting the merits and drawbacks of each method.

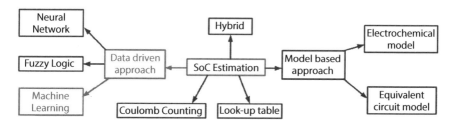

Figure 6.3 Classification of SoC estimation techniques.

6.2.1 Coulomb Counting Approach

Because of its simplicity, the Coulomb counting method is the most often used SoC estimating method. The SoC (SoC_t) at a particular time, t is estimated by considering the initial SoC (SoC_i) and an average of the discharging current (i_d) over the cycle as in (6.2). This method is also named as ampere-hour integral method as it is measured as current over a certain time period.

$$SoC_t = SoC_i + \frac{\eta}{Q_{rated}} \int_i^t i_d dt \qquad (6.2)$$

where η is the cell's coulombic efficiency.

The precision of the current sensors, as well as the initial SoC estimation, are key factors affecting the method's performance. In addition, the rated capacity must be recalculated from time to time to guarantee that the battery's maximum capacity is maintained. However, the shortcomings can be overcome by combining this method with other complex methods for better accuracy [19–21].

6.2.2 Look-Up Table Method

The external characteristic parameters of the battery cell like open-circuit voltage or discharging current, impedance, etc., highly influence the SoC estimation. Therefore, a direct mapping table can be generated by considering rigorous observations. The applications of this method are limited to laboratory uses only. The open-circuit voltage (V_{oc}) is assessed at regular intervals by removing the supply and allowing the battery to rest to obtain accurate results. As a result, real-world applications like transportation and grid integration are challenging to deploy.

6.2.3 Model-Based Methods

The model-based SoC estimation techniques are believed to be the most advanced and accurate way of estimation techniques as they consider the internal characteristics and dynamics of the batteries. They are broadly classified into two categories as electrochemical and equivalent circuit models.

6.2.3.1 Electrochemical Model

The electrochemical models are highly influenced by the internal chemical characteristics like kinetic process, the concentration of the electrolytes, the charging/discharging process, and potential distribution between the electrodes [8]. Due to the involvement of intensive domain knowledge, these methods provide accurate results. However, the same advantage also turns out to be the drawback of this method as it is difficult to realize all the parameters affecting chemically to determine the SoC.

6.2.3.2 Equivalent Circuit Model

To design an equivalent circuit model, the battery may also be described in terms of electrical components such as resistors, capacitors, and voltage sources. This is an approximate model that is inexpensive and effective for analysis. The most commonly employed models are Rint model, Thevenin model, and DP model [22]. The SoC is a non-linear function of the open-circuit voltage (V_{oc}) that is affected by the circuit configuration.

$$SoC = f(V_{oc}) \qquad (6.3)$$

For an instance, a modified Thevenin equivalent model is modeled by designing resistors, and capacitors representing the diffusion and polarization of the battery cell [23]. Figure 6.4 reflects the battery equivalent model considering the internal resistance (R_i), diffusion parameters (R_d, C_d), and polarization parameters (R_p, C_p). The external parameters of the battery like terminal voltage (V_{bt}), and battery current (i_{bt}) also influence the SoC estimation.

The adaptive filtration techniques like Kalman filters and their derivatives are utilized to assess SoC using model-based techniques [24–26].

6.2.4 Data-Driven Methods

The SoC is estimated using data-driven approaches by creating a supervised link between external characteristics such as terminal voltage of

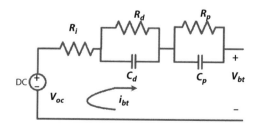

Figure 6.4 Thevenin equivalent model.

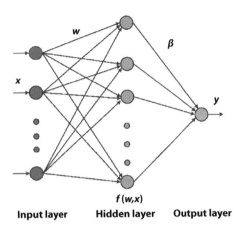

Figure 6.5 Basic structure of ANN.

the battery, its discharge current, temperature, and the battery's SoC. The supervised relationship is based on the non-linear mapping of the inputs with the instantaneous SoC. This technique overcomes the coulomb counting method's inaccuracy, the look-up table method's impracticability, and the complexity of model-based techniques. Examples of a few supervised learning techniques are artificial neural network (ANN), Fuzzy logic controller (FLC), machine learning (ML), and deep learning (DL). In Figure 6.5, the basic architecture of an ANN is shown, with x and y representing the input and output, respectively, and w representing the weight associated with the input and hidden layer.

The learning is independent of human supervision which reduces error and increases applicability. There have been developments in variants of ANN that differ in structure yet have similar functionalities like backpropagation network, recurrent network, and radial basis functional network [27–29].

6.2.5 Machine Learning–Based Methods

6.2.5.1 Support Vector Regression

One of the most widely applied machine learning approaches for classification applications is the support vector machine. It can also be utilized for solving regression problems as it shows good performance for prediction analysis. Hence, it is named support vector regression (SVR). The basic aim of the SVR technique is to define a hyperplane such that maximum data points are present on the hyperplane and minimize the upper bound of the generalization problem such that most of the data points stay within the boundaries. The data points close to the hyperplane are called support vectors [30]. The basic idea governing SVR is explained in Figure 6.6.

The linear regression function is given as:

$$y = f(x) = wx + b \tag{6.4}$$

where 'x' denotes the input, 'b' denotes the regression constant and 'w' indicates the weight coefficient.

In general, real-life problems are non-linear and cannot be linearly regressed. Therefore, the SVR technique maps the input into some non-linear functions for better prediction. The SVR function is generalized as:

$$y = f(x) = w \cdot \langle x, x_i \rangle + b \tag{6.5}$$

where $\langle x, x_i \rangle$ represents the mapping function, and the weight coefficient, 'w' can be determined by minimizing the weight function as:

$$\text{Minimize} \frac{1}{2}\|w\|^2 + c\sum_{i=1}^{N} \zeta_i + \zeta_i^* \tag{6.6}$$

Subjected to

$$y_i - wx_i - b \leq \varepsilon + \zeta_i$$
$$wx_i + b - y_i \leq \varepsilon + \zeta_i^* \tag{6.7}$$
$$\zeta_i, \zeta_i^* \geq 0$$

where ζ_i, ζ_i^* represents the distance between actual values outside the boundaries (slack variable) and corresponding boundary values (ε) of the tube.

It is essential to perform regression in higher dimensional feature space using an appropriate mapping function. Different types of kernel functions are employed to transfer low-dimensional input space into high-dimensional space, aiding regression speed.

$$\text{Linear kernel function: } K_L(x, x_i) = x^T x_i \tag{6.8}$$

$$\text{Polynomial kernel function: } K_p(x, x_i) = (1 + (x^T x_i)/a)^2 \tag{6.9}$$

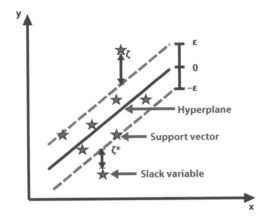

Figure 6.6 Basic structure of SVR.

Gaussian kernel function: $K_g(x, x_i) = exp(-||(x - x_i)||^2/2b^2)$ (6.10)

where a and b are the constants chosen suitably for accurate regression. The linear kernel function is easy to implement but is time-consuming. The polynomial kernel function of degree 2 is also called a quadratic kernel function. The Gaussian kernel function has a bell shape which leads to asymptotic accuracy but it is quite complex.

The Lagrangian multipliers in the Karush-Kuhn-Tucker conditions are used to solve the constrained optimization problem:

$$w^* = \sum_{i=1}^{N}(\lambda_i - \lambda_i^*) \cdot K(x, x_i)$$ (6.11)

where λ_i, λ_i^* are the lagrangian multipliers.

Hence, the regression function can be equated as:

$$y = \sum_{i=1}^{N}(\lambda_i - \lambda_i^*) \cdot K(x, x_i) + b$$ (6.12)

where $K(x, x_i)$ represents the kernel function mapping the two vectors x and x_i in its feature space.

6.2.5.2 Ridged Extreme Learning Machine (RELM)

Extreme learning machine belongs to the class of single-layer feed-forward neural network in which only one hidden layer is designed and the weights

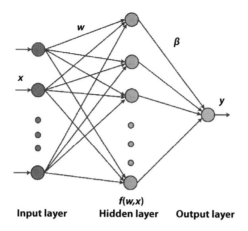

Figure 6.7 Structure of ELM.

associated with the input and hidden layer are randomly assigned [31]. Subsequently, the output can be calculated analytically. Figure 6.7 shows the architecture of basic ELM.

For a set of samples $\{S=(x_i,y_i)\}$, x_i represents the input and y_i denotes the output node, and $i=1,2..n$ where n is the no. of samples.

The output layer matrix is calculated as:

$$Y = H\beta \tag{6.13}$$

where H_l represents the hidden layer matrix and β denotes the weight associated with the output and the hidden layer.

$$H_l = \begin{bmatrix} e(x_i) \\ \vdots \\ e(x_N) \end{bmatrix} = \begin{bmatrix} e_1(x_1) & \cdots & e_L(x_1) \\ \vdots & \ddots & \vdots \\ e_1(x_N) & \cdots & e_L(x_N) \end{bmatrix}_{N \times L} \tag{6.14}$$

$$e(x) = F(w^T x) \tag{6.15}$$

where, $F(.)$ is the activation function like sigmoid, tanh, sin, cosine, radial, basis; w denotes the weight associated with input and hidden layer and L represents the no. of hidden layer node.

Since the hidden layer matrix (H_l) is not a square matrix; the output weight β is computed from equation (6.13) as:

$$\beta = H_l^\dagger Y \tag{6.16}$$

where, $H_l^\dagger$ represents the pseudo inverse of matrix H_l calculated by Moore Penrose inverse.

Generally, the small value of output weight (β) gives better generalization performance. Therefore, ridged regression is performed to minimize β using a suitable penalty factor as:

$$f(\beta) = \|H_l\beta - Y\|^2 + \frac{1}{R}\|\beta\|^2 \tag{6.17}$$

Differentiating (6.17) with respect to β and equating to zero, we get,

$$\frac{\partial f(\beta)}{\partial \beta} = -2H_l^T(Y-\beta) + \frac{2\beta}{R} = 0$$

Or

$$\Rightarrow 2H_l^T H_l \beta - 2H_l^T Y + \frac{2\beta}{R} = 0$$

Or

$$\Rightarrow \beta = H^T \left(H_l^T H_l + \frac{I}{R}\right)^{-1} Y \tag{6.18}$$

where $(1/R)$ is the penalty factor and the value varies between 2^{-22} and 2^{22}.

The solution obtained with $\left(\dfrac{I}{R}\right)$ term results in non-singular term $\left(H_l^T H_l + \dfrac{I}{R}\right)$ that achieves better generalization in performance [32]. The output node can be formulated as:

$$f(x) = \sum_{j=1}^{L} \beta_j [e(x)] = [e(x)]\beta = [e(x)]H_l^T \left(H_l^T H_l + \frac{I}{R}\right)^{-1} Y \tag{6.19}$$

The hyper-parameter like weights associated with hidden layer nodes and input nodes (w) are randomly generated which affects the overall performance.

6.3 BESS Description

A lithium-ion battery of model LG 18650HG2 is taken into consideration for the SoC estimation. The detailed configuration of the battery cell whose data are considered is listed in Table 6.1. The dataset comprises the charge-discharge cycle of the battery cell at a 1C rate for different temperature conditions [33]. The SoC is assessed by dividing the Ah data by the battery nominal capacity as in (6.1) considering discharge starting at 100% and charge ending at 100%.

6.4 Results and Discussion

The normalized data set of the drive cycle of the battery cell at two distinct temperatures (25°C and 10°C) is considered for analysis. The original cell output voltage, discharge current, and SoC are shown in Figure 6.8. Data normalization is performed to eliminate data redundancy effect while considering data with large values. The normalized data is calculated in the range of [-1, 1] as

$$d_{norm} = \frac{d - d_{min}}{d_{max} - d_{min}} \quad (6.20)$$

where d represents the original dataset, d_{norm} denotes the normalized dataset, d_{max} and d_{min} reperesnts the highest and lowest term in the dataset.

Table 6.1 Battery configuration.

Parameters	Units	Specifications
Cell Chemistry	-	Li[NiMnCo]O$_2$ (H-NMC)/Graphite+SiO
Nominal cell Voltage	V	3.6
Charge	A	1.5 (CC-CV) Normal
Discharge	V	2
Nominal cell Capacity	Ah	3.0
Cell energy Density	Wh/Kg	240

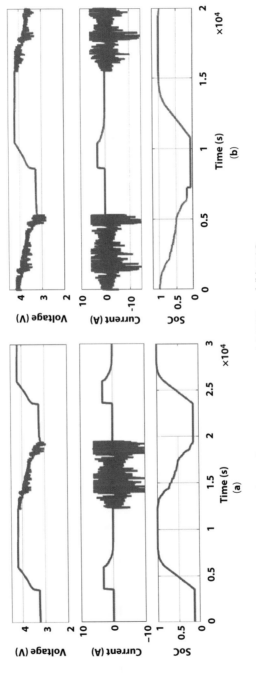

Figure 6.8 Original voltage, current, and SoC dataset at temperature (a) 25°C, and (b) 10°C.

The SoC is estimated in Matlab coding for the methods discussed in the earlier section. Considering the memory restrictions, 30,000 and 20,000 data at 25°C and 10°C correspondingly are considered. Of the data, 70% are trained and the remaining 30% of data are tested for validation of results. The performance measures like mean absolute error (MAE), root mean square error (RMSE), and mean square error (MSE) are well-defined in (6.21), (6.22), and (6.23) respectively, to measure the accuracy of estimation.

i. MAE: It is calculated as the mean absolute difference of the forecasted value (f_i) and the true value (t_i) of the dataset over the testing samples count (N_{test}). The lesser the value of MAE, the greater the prediction accuracy.

$$MAE = \frac{1}{N_{test}} \sum_{i=1}^{N_{test}} |t_i - f_i| \qquad (6.21)$$

ii. RMSE: It is another index to indicate the model prediction accuracy. It is a measure of the standard deviation of the estimated error.

$$RMSE = \sqrt{\frac{1}{N_{test}} \sum_{i=1}^{N_{test}} (t_i - f_i)^2} \qquad (6.22)$$

iii. MSE: It is a measure of the quality of the prediction algorithm and is calculated as the average squared error. Squaring the error eliminates the negative terms and further increases the large terms that give the user an idea to reduce those values for a better estimate.

$$MSE = \frac{1}{N_{test}} \sum_{i=1}^{N_{test}} (t_i - f_i)^2 \qquad (6.23)$$

The comparative analysis of performances of SVR with different kernel functions, ELM, and ridged-regression-based ELM (RELM) is presented in Figure 6.9 at 25°C and 10°C. Table 6.2 outlines the error parameters enumerating various approaches. The graphical representations of the performance indices give clarity on the superior performance of the proposed RELM technique as in Figure 6.10.

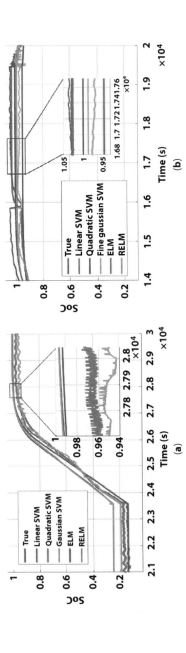

Figure 6.9 SoC estimation using linear SVR, quadratic SVR, Gaussian SVR, ELM, and RELM at (a) 25°C, and (b) 10°C.

Table 6.2 Performance evaluation of different methods based on performance indices.

Methods	Parameters					
	RMSE		MAE		MSE	
	25°C	10°C	25°C	10°C	25°C	10°C
Linear SVM	0.029	0.031	0.025	0.026	0.00087	0.00095
Quadratic SVM	0.036	0.024	0.033	0.021	0.00135	0.00060
Gaussian SVM	0.048	0.029	0.044	0.026	0.00229	0.00085
ELM	0.0158	0.0145	0.0124	0.0135	0.000248	0.00021
RELM	0.0094	0.0061	0.0029	0.0057	0.000088	0.000037

6.5 Conclusion

An energy management system is an important aspect of microgrid control. This chapter highlights the management of BES by evaluating SoC which is crucial for optimum battery life. The irregularity of the battery's internal chemical properties influences the battery parameters that determine SoC. This creates the need for a data-driven technique for SoC assessment. The conventional data-driven techniques like ANN and FLC are time-consuming and are difficult for practical implementation. Therefore, machine learning techniques are introduced that improve predictability and accuracy. This chapter compares and contrasts various machine learning algorithms for predicting battery SoC. Different kernel functions–based SVR techniques are implemented to estimate the SoC of the lithium-ion battery at two different temperatures. The major drawbacks of the SVR algorithm are inaccuracy for large datasets, time-consuming, difficulty in coefficient estimations, and complexity in calculations. These shortcomings of the SVR technique are overcome by the ELM technique. Furthermore, the output weight becomes minimum by adding the ridged regression analysis in the ELM that addresses the generalization issue with ultimate precision. The efficacy of the proposed regression-based ELM technique is observed with regard to the performance indices.

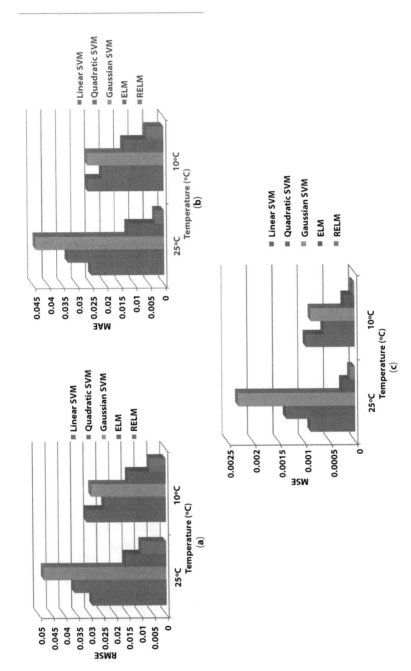

Figure 6.10 Graphical representation of the comparative analysis based on (a) RMSE, (b) MAE, and (c) MSE.

References

1. https://www.iea.org/reports/global-energy-review-2021/co2-emissions
2. U. Datta, A. Kalam, and J. Shi, "A review of key functionalities of battery energy storage system in renewable energy integrated power systems," *Energy Storage*, vol. 3, no. 5, pp. 224, 2021.
3. M. Farhadi, and O. Mohammed, "Energy storage technologies for high-power applications," *IEEE Transactions on Industry Applications*, vol. 52, no. 3, pp. 1953-1961, 2015.
4. D. Akinyele, J. Belikov, and Y. Levron, "Battery storage technologies for electrical applications: Impact in stand-alone photovoltaic systems," *Energies*, vol. 10, no. 11, pp. 1760, 2017.
5. M. Sufyan, N. A. Rahim, MM. Aman, C. K. Tan, and S. R. S. Raihan, "Sizing and applications of battery energy storage technologies in smart grid system: A review," *Journal of Renewable and Sustainable Energy*, vol. 11, no. 1, pp. 014105, 2019.
6. B. Diouf, and R. Pode, "Potential of lithium-ion batteries in renewable energy," *Renewable Energy*, vol. 76, pp. 375-380, 2015.
7. A. Kumar, Y. Deng, X. He, P.Kumar, and R.C.Bansal, "Energy management system controller for a rural microgrid," *Journal of Engineering*, vol. 13, pp. 834-839, 2017.
8. M. T. Lawder, B. Suthar, P. W C. Northrop, S. De, C. M. Hoff, O. Leitermann, M. L. Crow, S. S. Shriram, and V. R. Subramanian, "Battery energy storage system (BESS) and battery management system (BMS) for grid-scale applications," *Proceedings of the IEEE*, vol. 102, no. 6, pp. 1014-1030, 2014.
9. R. Xiong, J. Cao, Q. Yu, H. He, and F. Sun, "Critical review on the battery state of charge estimation methods for electric vehicles," *IEEE Access*, vol. 6, pp. 1832-1843, 2017.
10. C. Lin, A. Tang, and J. Xing, "Evaluation of electrochemical models based battery state-of-charge estimation approaches for electric vehicles," *Applied Energy*, vol. 207, pp. 394-404, 2017.
11. H. He, R. Xiong, and J. Fan, "Evaluation of lithium-ion battery equivalent circuit models for state of charge estimation by an experimental approach," *Energies*, vol. 4, no. 4, pp. 582-598, 2011.
12. D. NT. How, MA. Hannan, MS. H. Lipu, and P.J. Ker, "State of charge estimation for lithium-ion batteries using model-based and data-driven methods: A review," *IEEE Access*, vol. 7, pp. 136116-136136, 2019.
13. D. Saji, P. S. Babu, and K. Ilango, "SoC Estimation of Lithium Ion Battery Using Combined Coulomb Counting and Fuzzy Logic Method," *2019 4th International Conference on Recent Trends on Electronics, Information, Communication & Technology (RTEICT)*, pp. 948-952, 2019.
14. M. S. Sidhu, D. Ronanki, and S. Williamson, "Hybrid state of charge estimation approach for lithium-ion batteries using k-nearest neighbour

and gaussian filter-based error cancellation," *2019 IEEE 28th International Symposium on Industrial Electronics (ISIE)*, pp. 1506-1511, 2019.
15. M. S. H. Lipu, M. A. Hannan, A. Hussain, M. H. Saad, A. Ayob, and M. N. Uddin, "Extreme learning machine model for state-of-charge estimation of lithium-ion battery using gravitational search algorithm," *IEEE Transactions on Industry Applications*, vol. 55, no. 4, pp. 4225-4234, 2019.
16. T. Hansen, and C. J. Wang, "Support vector based battery state of charge estimator," *Journal of Power Sources*, vol. 141, no. 2, pp. 351-358, 2011.
17. J. C. Álvarez Antón, P. J. García Nieto, C. Blanco Viejo, and J. A. Vilán Vilán, "Support Vector Machines Used to Estimate the Battery State of Charge," *IEEE Transactions on Power Electronics*, vol. 28, no. 12, pp. 5919-5926, 2013.
18. Z. Wang, and D. Yang, "State-of-charge estimation of lithium iron phosphate battery using extreme learning machine," *2015 6th International Conference on Power Electronics Systems and Applications (PESA)*, pp. 1-5, 2015.
19. M. A. Awadallah, and B. Venkatesh, "Accuracy improvement of SoC estimation in lithium-ion batteries," *Journal of Energy Storage*, vol. 6, pp. 95-104, 2016.
20. F. Codeca, S. M. Savaresi, and G. Rizzoni, "On battery state of charge estimation: A new mixed algorithm," *2008 IEEE International Conference on Control Applications*, pp. 102-107, 2008.
21. K. V. Singh, H. O. Bansal, and D. Singh, "Hardware-in-the-loop implementation of ANFIS based adaptive SoC estimation of lithium-ion battery for hybrid vehicle applications," *Journal of Energy Storage*, vol. 27, pp. 101124, 2020.
22. H. He, R. Xiong, H. Guo, and S. Li, "Comparison study on the battery models used for the energy management of batteries in electric vehicles," *Energy Conversion and Management*, vol. 64, pp. 113-121, 2012.
23. J. P. Rivera-Barrera, M. G. Nicolás, and H. O. Sarmiento-Maldonado, "SoC estimation for lithium-ion batteries: Review and future challenges," *Electronics*, vol. 6, no. 4, pp. 102, 2017.
24. H. He, R. Xiong, X. Zhang, F. Sun, and J. Fan, "State-of-charge estimation of the lithium-ion battery using an adaptive extended Kalman filter based on an improved Thevenin model," *IEEE Transactions on Vehicular Technology*, vol. 60, no. 4, pp. 1461-1469, 2011.
25. S. Wang, C. Fernandez, L. Shang, Z. Li, and J. Li, "Online state of charge estimation for the aerial lithium-ion battery packs based on the improved extended Kalman filter method," *Journal of Energy Storage*, vol. 9, pp. 69-83, 2017.
26. C. Zhang, K. Li, L. Pei, and C. Zhu, "An integrated approach for real-time model-based state-of-charge estimation of lithium-ion batteries," *Journal of Power Sources*, vol. 283, pp. 24-36, 2015.
27. M. Ismail, R. Dlyma, A. Elrakaybi, and R. Ahmed, and S. Habibi, "Battery state of charge estimation using an Artificial Neural Network," *2017 IEEE*

Transportation Electrification Conference and Expo (ITEC), pp. 342-349, 2017.
28. F. Yang, W. Li, C. Li, and Q. Miao, "State-of-charge estimation of lithium-ion batteries based on gated recurrent neural network," *Energy*, vol. 175, pp. 66-75, 2019.
29. H. F. Kong, L. J. Xiang, and S. L. Xia, "SoC Estimation on Radial Basis Function," *Advanced Materials Research*, vol. 403, pp. 3119-3122, 2012.
30. A. L. Smola, and B. Schölkopf, "A tutorial on support vector regression," *Statistics and Computing*, vol. 14, no. 3, pp. 199-222, 2004.
31. G. B. Huang, Q. Y. Zhu, and C. K. Siew, "Extreme learning machine: theory and applications," *Neurocomputing*, vol. 70, no.1-3, 489-501, 2006.
32. G. B. Huang, H. Zhou, X. Ding, and R. Zhang, "Extreme learning machine for regression and multiclass classification," *IEEE Transactions on Systems, Man, and Cybernetics, Part B (Cybernetics)*, vol. 42, no. 2, pp. 513-529, 2011.
33. P. Kollmeyer, C. Vidal, M. Naguib, and M. Skells, "LG 18650HG2 Li-ion Battery Data and Example Deep Neural Network xEV SoC Estimator Script", Mendeley Data, V3, 2020.

7
Dual-Energy Storage System for Optimal Operation of Grid-Connected Microgrid System

Deepak Kumar[1*] **and Sandeep Dhundhara**[2]

[1]Department of Electrical and Electronics Engineering, UIET, Panjab University, Chandigarh, India
[2]Department of Basic Engineering, COAE&T, CCS Haryana Agricultural University, Hisar, India

Abstract

The advances in renewable power generation technologies and modernization of power systems allow the growing proliferation of renewable energy resources (RERs) in the form of microgrid (MG). The integration of microgrids in the distribution system changes the passive distribution system into an active distribution system. The uncertainty in power generation from RERs is causing the major operational challenge to meet supply-load demand balance. Therefore, the electrical energy storage system (ESS) has become an important component of the microgrid system. The energy storage operating time limits have a great impact on the operating cost as well as on the life cycle of the storage. In this research work, the dual energy storage system (DESS) including battery storage (BS) and pump hydro storage (PHS) has been investigated to understand the impact of the minimum operating time limit on the optimal operating cost. Three operating cases have been studied to obtain the optimal operating cost of microgrid system operating with single energy storage and with DESS for different operating time constraints. The optimal operating time increases the effective life cycle of the storage and optimizes the operating cost very effectively. General algebraic mathematical systems (GAMS) modeling for mixed integer nonlinear optimization problem has been developed and simulated using the discrete and continuous optimizer (DICOPT) solver. The results show that the optimal minimum operating time constraint has a great impact on the optimal operating cost of the system.

---

*Corresponding author: dk_uiet@pu.ac.in

Sandeep Dhundhara, Yajvender Pal Verma, and Ashwani Kumar (eds.) Energy Storage Technologies in Grid Modernization, (181–212) © 2023 Scrivener Publishing LLC

Keywords: Microgrid, renewable energy resources, pump hydro storage, battery storage, dual-energy storage

7.1 Introduction

The global energy demand has been increased drastically in the last few decades due to modernization and technological advancements in the existing power systems to improve reliability and economic growth. The major share of the power comes from fossil fuels to fulfill the power requirement of the system among all available resources either renewable or conventional. The proper and planned minimum utilization of these fossil fuels leads to a reduction in environmental and global warming concerns [1]. At the same time, dependency on fossil fuels has been reduced by integrating the green renewable energy resources (RERs) and small-scale distributed generations (DGs) like solar photovoltaic (PV), wind turbine (WD), small hydro generators, diesel generators, biogas, geothermal, fuel cell, etc. The variability in power demand–generation balance caused by intermittent resources can effectively countered by the energy storage system (ESS) as an essential part of modern power system [2]. Otherwise, the intermittent nature of the renewable resources brings the system power balance into question and leads to unreliable operation [3]. Increase penetration of these intermittent renewable resources leads to frequent undesired power fluctuations which need storage systems that are capable of managing the system power balance. The battery storage (BS) and pump hydro storage (PHS) can complement each other for different power fluctuations in the system. The dual energy storage system (DESS) is better than the individual storage system as it possesses the advantages of both storage systems. The system stability and realibity under question due to frequent variations in generation and demand can be managed by controlling both demand and generation.

The deregulation and technological innovations in modern power system allow the participation of different scaled distributed generators working in a group known as a microgrid (MG) with capabilities to operate either in islanding (IM) or grid-connecting mode (GCM) [4, 5]. The microgrid operating with the proper storage provision can manage the variability of the available intermittent resources very effectively [6]. The microgrid includes small-scale schedulable and non-schedulable power generators, electrical loads (fixed and adjustable or a combination of both), energy storage systems (single or multiple), and a controller. In GCM, the microgrid manages the variability through power exchanges with storage (if available) and the utility grid. Energy storage of the MG can effectively counter the variability of RERs by scheduling its charging or discharging modes whenever power stress occurs.

Pump hydro and battery storage systems are highly utilized ESSs among many available in modern power systems to manage the system variability and to obtain the optimal operating cost. Due to inherent properties the battery and pump hydro based ESSs are most suitable for respectively small power and large power variations [7]. Therefore, storage provides an effective operation of MG system having large integration of the intermittent resources. The battery storage system requires additional power electronics devices for power conversion as compared to the pump hydro storage system to make it compatible with the utility grid. Therefore, the operating cost of the battery storage is higher than the pump hydro storage [8]. It is crucial to select an effective and suitable storage system for the microgrid operation as both storage systems have many advantages and disadvantages along with different operating costs. The performance of the MG system during power stress periods can be improved effectively by using a combined battery and pump hydro based energy storage systems in the form of DESS [9]. The DESS manages the small frequent and large slow power variations effectively through the battery and pump hydro based storage units, respectively. Therefore, a joint operation of the pump hydro and battery storage in a microgrid system can manage the power fluctuations either due to power generation or due to load demand effectively by complementing each other [10].

In past, significant studies have been carried out to address the impact of increasing integration of highly intermittent RERs in the existing power system along with controlling system variability [11, 12]. The large optimization problems of the power system have been solved with reliable solutions by implementing deterministic methods. A few deterministic optimizing scheduling of microgrid have been conducted and analyzed for grid-connected and islanding modes of operation in electricity markets [13–16]. The intermittency of the power generation from various RERs adversely affects the system reliability, power quality, and optimal operation in large. This sometimes leads to a locally optimal solution value rather than the global optimal solution for the microgrid operation. In [17–20] the microgrid issues have been studied by implementing stochastic and heuristic approaches. The latest studies addressing the energy management of microgrid systems in literature have been tabulated in Table 7.1 based on sources, type of storage, approach used, uncertainty in generation and application, etc., for optimal solutions.

Table 7.1 Literature summary of the present and previous studies.

Ref. no. (Year)	Microgrid resources and mode of operation	Number of energy storage	Objective to minimize	Consider uncertainty	Approach applied	Application
[21] (2022)	Solar, wind, CHHP, boiler and diesel generators Grid connected	One	Cost	Yes	Nonlinear programing	Energy management Bidding
[5] (2021)	Wind, solar, geothermal, hydro Grid connected	One	Cost	Yes	Nonlinear programing	Optimal sizing and placement
[22] (2021)	Solar Grid connected	One	Cost	Yes	Linear programing	Energy management
[23] (2021)	Solar and diesel generator Grid connected	One	Cost	Yes	LAPO and ABC	Optimal location and energy management
[24] (2021)	CHP, wind and boiler Grid connected	Three	Cost	No	Linear programing	Energy management

(Continued)

Table 7.1 Literature summary of the present and previous studies. *(Continued)*

Ref. no. (Year)	Microgrid resources and mode of operation	Number of energy storage	Objective to minimize	Consider uncertainty	Approach applied	Application
[1] (2020)	Solar, wind, fuel cell, microturbine and diesel generator Grid connected	One	Cost	Yes	Linear programing	Energy management
[9] (2020)	Wind, solar, boiler, CHP and diesel generator Grid connected	One	Cost	Yes	Nonlinear programing	Energy management
[25] (2020)	Wind and diesel generator Grid and islanded	One	Cost	Yes	Linear programing	Energy management
[26] (2020)	Wind, solar, diesel, boiler and CHP Grid connected	One	Cost	Yes	Nonlinear programing	Energy management
[27] (2019)	Solar, wind and diesel Grid connected	Zero	Cost	Yes	Nonlinear programing	Energy management
[28] (2019)	Solar, wind and micro-turbine Islanded	One	Cost	Yes	TLBO algorithm	Energy management

(Continued)

Table 7.1 Literature summary of the present and previous studies. (*Continued*)

Ref. no. (Year)	Microgrid resources and mode of operation	Number of energy storage	Objective to minimize	Consider uncertainty	Approach applied	Application
[29] (2018)	Wind Grid conected	One	Cost	Yes	Nonlinear programing	Energy management
[30] (2018)	CHP, wind and bolier Islanded	One	Cost	Yes	Linear programing	Energy management
[31] (2017)	Wind, solar thermal and pump hydro Grid connected	Zero	Cost	Yes	Nonlinear programing	Energy management
[32] (2017)	Solar, wind, fuel cell, micro-turbine and diesel Grid connected	One	Cost	No	Linear programing	Energy management and sizing
[33] (2016)	Wind, diesel, fuel cell, solar and microturbine Grid connected	One	Cost	Yes	--	Energy management
Current Study	Wind, solar, CHP, boiler and diesel Grid connected	Two	Cost	Yes	Nonlinear programing	Energy management

lightning attachment procedure optimization (LAPO) and artificial bee colony (ABC), teaching-learning-based optimization (TLBO).

In literature, microgrid either in grid-connected mode [1, 5, 24, 26, 34] or islanded mode [25, 28, 30] of operation have been investigated considering different RERs and storage systems. The main objective of the present research is to optimize the overall operational cost of the microgrid system. It has been found through reviewing the existing literature that appreciable work has been done in considering energy management for improving system efficiency, reliability, and effective participation in electricity markets. The cost-effectiveness of individual storage systems has been investigated for managing the system power variability caused by load demand and intermittent resources. However, the impact of the minimum on and off time of the DESS has not been investigated for the GCMG system.

Present study, focuses on the use of dual energy storage system to manage the system variability associated with the intermittent solar and wind plants and compare it with results of MG operating with either battery storage or pump hydro storage as single storage. Three cases have been simulated as mixed integer nonlinear programming (MINLP) problem incorporating different technical and operational constraints on the system components. The MG system has a limited power exchangeable capacity with the grid depending mainly on the transmission line capacity. The impact of minimum on and off time limits of the individual and dual storage system on the system's operational cost. Due to the nature of the technical and operational constraints of the system, this optimization problem becomes a non-linear problem.

The promising features of the present proposed work are summarized as follows:

1. Optimal operating cost solution has been proposed for GCMG system incorporating the battery storage and pump hydro storage as a dual energy storage.
2. Mathematical equation modelling of shiftable and controllable loads considering minimum time of operation, and total energy consumed to manage the supply-demand effectively.
3. The impact of minimum on and off time of the storage system of single and dual ESS on the overall operational cost has also been evaluated for the MG system.

The simulated results obtained and presented in the present research work evaluate and analyze the impact of the minimum on and off-time scheduling on the operational cost of the MG system for managing the system power supply-demand balance.

The rest of the chapter is arranged into four sections. Section 7.2 elaborates the mathematical equations used for the modeling of diesel generator, PV, WD, BS, PHS, and adjustable loads based on operational and technical constraints of the MG system operating with limited power exchange capabilities. The problem formulation for analyzing the impact of the minimum operating time constraint of the storage system on the optimal cost of operation under investigation has been explained in Section 7.3. Section 7.4 explores the system data, simulation results, and explanations of the MG system under study and section 7.5 finally concludes the study.

7.2 System Mathematical Modelling

The grid-connected microgrid system under this study consists of diesel generators, wind turbine power plants, solar plant, adjustable loads, fixed load, and the DESS (battery and the pump hydro). This section expresses the various components and the processes adopted in mathematical equation form operation of grid-connected MG system in an electricity market operating with a DESS including battery storage and pump hydro storage for optimal operation. The ESS not only manages the variability of the intermittent resources but also gives confidence to the market player to participate in the electricity market. The storage cannot be operative all the time as they need the minimum charging time to be able for discharging. Therefore, the minimum charging and discharging time requirement of the energy storage system becomes an important constraint to be explored. To demonstrate the impact of minimum charging/motoring and discharging/generating modes of the DESS on the optimal operation of the GCMG system has been developed and analyzed using a MINLP subject to various technical and operational constraints. The power demand and generation are both dynamic in nature and therefore power exchange between the grid and the microgrid has been considered through a constraint. The system operator manages the power exchange limit and plans the scheduling of resources one day-ahead of actual scheduling knowing the predicted patterns of generation, load demand, and energy prices. The DESS along with the power exchange option has been applied to minimize the cost of GCMG operation. Various components of the MG system have been described in mathematical equations as below.

7.2.1 Modelling of Wind Turbine Power Generator

The wind carries energy and can be converted into a suitable form of energy using wind turbines. A wind turbine converts the mechanical energy associated with the wind into electrical energy. The total power generated from the wind turbine at t^{th} the time interval can be mathematically represented as (7.1).

$$P_{gen\,wind(t)} = 0.5\,\eta_{wind}\,\rho_{wind(t)}\,C_{wind}\,A_{rotor}\,V_{wind(t)}^{3} \qquad (7.1)$$

Where, $P_{gen\,wind(t)}$ is the power generated by the wind turbine power generator at t^{th} time interval, $\rho_{wind(t)}$ is the air density of the wind at t^{th} time interval, $V_{wind(t)}$ is the velocity of the wind at t^{th} time interval, A_{rotor} is the swept area of the wind turbine blades, η_{wind} is the efficiency of the wind turbine and C_{wind} is power coefficient of the wind turbine [35, 36].

7.2.2 Modelling of Solar Power Plant

The generated power of the PV array is highly intermittent and varies with the environmental conditions. The power generation at t^{th} the time interval can be mathematically expressed as (7.2).

$$P_{gen\,solar(t)} = \eta_{solar}\,A_{Csolar}\,I_{Rsolar(t)} \qquad (7.2)$$

Where, $P_{gen\,solar(t)}$ is the power generated by the solar plant at t^{th} time interval, A_{Csolar} is the area of the solar array and $I_{Rsolar(t)}$ is the solar irradiation at t^{th} time interval η_{solar} is the efficiency of the solar unit/plant [35, 36].

7.2.3 Modelling of Conventional Diesel Power Generator

The diesel generator in a microgrid system produces power during the time intervals when the other distributed generators are not able to meet the system demand due to intermittency in the renewable power generation from renewable resources like solar and wind along with the load demand. The fuel cost of the generator is governed by the fuel cost coefficients associated with the respective generator and can be expressed as a quadratic function of the power generated (7.3).

$$C_{fuel\,gen(i,t)} = a_i P_{gen(i,t)}^2 + b_i P_{gen(i,t)} + c_i \qquad (7.3)$$

Where, $P_{gen(i,t)}$ is the power generated from the ith diesel generator at t^{th} time interval, and a_p, b_p, and c_p are the fuel coefficients of the ith diesel generator, respectively.

7.2.4 Modelling of Combined Heat and Power (CHP) and Boiler Plant

The mathematical modelling of the cost of power generation from the combined heat and power plant has been considered similar to the diesel generator and can be expressed as the quadratic function of the power generated and the fuel coefficients as in (7.4). The boiler plant is modelled as a linear function of the power generated as (7.5)

$$C_{chp(j,t)} = a_j P_{chp(j,t)}^2 + b_j P_{chp(j,t)} + c_j \qquad (7.4)$$

$$C_{br(k,t)} = b_k P_{br(k,t)} \qquad (7.5)$$

Where, $C_{chp(j,t)}$ and $C_{br(k,t)}$ are the fuel cost of the combined heat and power plant and the boiler plant respectively for the jth CHP and the kth boiler at t^{th} time interval. $P_{chp(j,t)}$ and $P_{br(k,t)}$ are the power generation from the CHP plant and boiler plant, respectively.

7.2.5 Modelling of Dual Energy Storage System

ESS become an integral part of the microgrid system due to the presence of solar and wind renewable intermittent power resources. The principal function of the energy storage system is to manage the power balance by operating in charging or discharging modes whenever put under stress either due to fluctuation in generation or demand or a combination of generation and demand. The battery storage system can manage the frequent small power variations of the system, whereas the pump hydro storage plant is suitable for managing slow variations of large magnitude. The dual energy storage system has the advantages of both the storage systems and the power variability of any nature can easily be managed. Both the energy storage systems have a similar operation.

7.2.5.1 Battery Bank Storage System

The battery bank storage system consists of the number of batteries. The state of charge of the battery bank storage system at t^{th} the time interval will be evaluated based on the energy stored or energy delivered at end of the charging or discharging intervals and expressed by SOC as in (7.6). The operating cost of the battery bank storage system is assumed to be fixed by the charging and discharging power and expressed as (7.7).

$$SOC_{bs(s,t)} = SOC_{bs(s,t-1)} - P_{bs\,dch(s,t)} m_{bs(s,t)} \Delta t / \eta_{bs\,dch} + P_{bs\,ch(s,t)} n_{bs(s,t)} \Delta t\, \eta_{bs\,ch} \quad (7.6)$$

$$C_{bs(s,t)} = h_{bs\,dchg} * P_{bs\,dch(s,t)} m_{bs(s,t)} + g_{bs\,chg} * P_{bs\,ch(s,t)} n_{bs(s,t)} \quad (7.7)$$

Where, $SOC_{bs(s,t)}$ is the state of charge of the sth battery bank at t^{th} time interval, $\eta_{bs\,dch}$ and $\eta_{bs\,ch}$ are the efficiency of battery bank during discharging and charging mode of operation respectively, $m_{bs(s,t)}$ and $n_{bs(s,t)}$ are the binary variable indicating discharging and charging modes, respectively. Δt is the minimum time interval of charging and discharging mode.

7.2.5.2 Pump Hydro Storage System

The pump hydro storage plant consists of a reservoir to store the water from other resources like rivers, lakes, and other man-made or natural reservoirs, etc. The state of the reservoir at t^{th} time interval will be evaluated based on the energy stored or delivered at end of the pumping or generating intervals and expressed by SOC_{ph} as in (7.8). The operating cost of the pump hydro storage system is assumed to be fixed by the motoring and generating power and expressed as (7.9).

$$SOC_{ph(p,t)} = SOC_{ph(p,t-1)} - P_{ph\,gen(p,t)} m_{ph(p,t)} \Delta t / \eta_{ph\,gen} + P_{ph\,mot(p,t)} n_{ph(p,t)} \Delta t\, \eta_{ph\,mot} \quad (7.8)$$

$$C_{ph(p,t)} = k_{ph\,gen} * P_{ph\,gen(p,t)} m_{ph(p,t)} + l_{ph\,mot} * P_{ph\,mot(p,t)} n_{ph(p,t)} \quad (7.9)$$

Where, $SOC_{ph(p,t)}$ is the state of the reservoir of the pth pump hydro plant at t^{th} time interval, $\eta_{ph\,gen}$ and $\eta_{ph\,mot}$ are the efficiency of pump hydro storage during generating and motoring mode of operation respectively, $m_{ph(p,t)}$ and $n_{ph(p,t)}$ are the binary variable indicating generating and motoring modes, respectively. Δt is the minimum time interval of charging and discharging mode.

7.2.6 Modelling of Power Transfer Capability

The presence of unpredictable renewable energy resources can effectively countered with power exchange option for the grid connected microgrid system to manage its power balance when it operates without energy storage system or with storage having limited capacity. It also helps in optimizing the operating cost of the system by trading the excessive power with the grid. The selling and purchasing power costs can be the same or different depending upon the service requirements. In this research, the selling power cost is 20 percent higher than the purchasing power cost considered a penalty factor to support the grid. The net power exchange cost is modeled as (7.10) [2, 37].

$$Cost_{Pexchange}(t) = \begin{cases} Penalty_{down} * Energy_{Price}(t) * P_{exchange}(t) & \text{if } P_{exchange}(t) > 0 \\ 0 & \text{if } P_{exchange}(t) = 0 \\ -Penalty_{up} * Energy_{Price}(t) * P_{exchange}(t) & \text{if } P_{exchange}(t) < 0 \end{cases}$$

(7.10)

7.3 Objective Function and Problem Formulations

Objective function (OF)
The coordination of all the resources, loads and dual-energy storage systems for the optimal operational cost of the system is the main objective of the problem subject to technical and operational constraints. The objective function is represented by (7.11) [2, 37, 38].

$$\min OF = \sum_{t=1}^{T}(\sum_{i=1}^{I} Cost_{Pgen(i,t)} + \sum_{i=1}^{J} Cost_{CHP(j,t)} + \sum_{i=1}^{K} Cost_{boiler(k,t)}$$
$$+ \sum_{r=1}^{R} Cost_{Pgen\,rew(r,t)} + Cost_{Pexchange(t)} + Cost_{bs\,ch\,and\,dch(t)} + Cost_{ph\,mot\,and\,gen(t)})$$

(7.11)

Constraints associated with power balance (7.12), power exchange between the grid and the microgrid (7.13), dispatchable DGs (7.14)-(7.18), battery storage (7.19)-(7.24), pump hydro storage (7.25)-(7.30), adjustable loads (7.31)-(7.33).

7.3.1 Operational and Technical Constraints

A practical system must fulfill certain operational constraints for a realiable and secure operation. All parameters must remain within the limitations set by the system operator in the form of constraints [2, 9, 27, 37–39].

Optimal Operation of GCMG with DESS   193

$$\sum_{i=1}^{I} P_{gen(i,t)} + \sum_{r=1}^{R} P_{gen\,rew(r,t)} + P_{exchange(t)} + P_{ph\,gen(p,t)} + P_{bs\,dch(s,t)}$$
$$= \sum_{d=1}^{D} P_{dem(d,t)} + P_{ph\,mot(p,t)} + P_{bs\,ch(s,t)} \tag{7.12}$$

$$-P_{exchange}^{max}(t) \leq P_{exchange}(t) \leq P_{exchanged}^{max}(t) \tag{7.13}$$

$$P_{gen(i,t)}^{min} \leq P_{gen(i,t)} \leq P_{gen(i,t)}^{max} \tag{7.14}$$

$$P_{gen(i,t)} - P_{gen(i,t-1)} \leq U_{p\,rate(i)} \tag{7.15}$$

$$P_{gen(i,t-1)} - P_{gen(i,t)} \leq D_{w\,rate(i)} \tag{7.16}$$

$$T_{on(i)} \geq U_{p\,time(i)} \left(I_{g(i,t)} - I_{g(i,t-1)} \right) \tag{7.17}$$

$$T_{off(i)} \geq D_{w\,time(i)} \left(I_{g(i,t)} - I_{g(i,t-1)} \right) \tag{7.18}$$

$$P_{bs(s,t)} \leq P_{bs\,dch(s,t)}^{max} m_{bs(s,t)} - P_{bs\,ch(s,t)}^{min} n_{bs(s,t)} \tag{7.19}$$

$$P_{bs(s,t)} \geq P_{bs\,dch(s,t)}^{min} m_{bs(s,t)} - P_{bs\,ch(s,t)}^{max} n_{bs(s,t)} \tag{7.20}$$

$$m_{bs(s,t)} + n_{bs(s,t)} \leq 1 \tag{7.21}$$

$$SOC_{bs\,min(s)} \leq SOC_{bs(s,t)} \leq SOC_{bs\,max(s)} \tag{7.22}$$

$$T_{ch(s,t)} \geq U_{P\,time(s)}^{min} \left(m_{bs(s,t)} - m_{bs(s,t-1)} \right) \tag{7.23}$$

$$T_{dch(s,t)} \geq D_{w\,time(s)}^{min} \left(n_{bs(s,t)} - n_{bs(s,t-1)} \right) \tag{7.24}$$

$$P_{ph(p,t)} \leq P_{ph\,gen(p,t)}^{max} m_{ph(p,t)} - P_{ph\,mot(p,t)}^{min} n_{ph(p,t)} \tag{7.25}$$

$$P_{ph(p,t)} \leq P_{ph\,gen(p,t)}^{min} m_{ph(p,t)} - P_{ph\,mot(p,t)}^{max} n_{ph(p,t)} \tag{7.26}$$

$$m_{ph(p,t)} + n_{ph(p,t)} \leq 1 \tag{7.27}$$

$$SOC_{ph\,min(p)} \leq SOC_{ph(p,t)} \leq SOC_{ph\,max(p)} \tag{7.28}$$

$$T_{ph\,mot(p,t)} \geq U_{P\,time(p)}^{min}\left(m_{ph(p,t)} - m_{ph(p,t-1)}\right) \tag{7.29}$$

$$T_{ph\,gen(p,t)} \geq D_{w\,time(p)}^{min}\left(n_{ph(p,t)} - n_{ph(p,t-1)}\right) \tag{7.30}$$

$$P_{dem(d)}^{min}\,q_{(d,t)} \leq P_{dem(d,t)} \leq P_{dem(d)}^{max}\,q_{(d,t)} \tag{7.31}$$

$$T_{dem(d)} \geq U_{op\,time(d)}^{min}\left(q_{(d,t)} - q_{(d,t-1)}\right) \tag{7.32}$$

$$\sum_{\alpha}^{\beta} P_{dem(d,t)} = E_{dem(d)} \tag{7.33}$$

The objective function considered in (7.11) minimizes the daily operational cost of microgrid, which includes the cost of power generation from the diesel generators, cost of power exchange, cost of charging-discharging power of battery storage, cost of motoring and generating of the pump hydro storage, cost of power generation from combined heat and power plant and the cost of power generation from the boiler unit. The power balance equation (7.12) of the system always ensures that the total sum of the power generation from the system resources, storage, and power exchange must match the system power demand. Depending upon the system requirements the power exchange between the microgrid and the grid can be positive (power purchased), negative (power sold), or zero. The power exchange is limited by the transmission line capacity connecting the microgrid with the grid (7.13). The constraint in (7.14) defines the power generation limits (maximum and minimum) of the dispatchable units at any time interval, and technical constraints (ramp rate up/down and minimum operation time on/off) are defined in (7.15)-(7.16) and (7.17)-(7.18), respectively, where I_g is the binary variable representing the commitment state of the dispatchable unit ($I_g = 1$ when unit is committed and $I_g = 0$ when unit is not committed). In (7.19) and (7.20) the charging and discharging power limits are defined on the battery storage unit based on the mode of operation. m_{bs} and n_{bs} are the binary variables indicating the mode of operation (m_{bs} = 1 means charging mode 0 otherwise and if n_{bs} = 1 means discharging mode, 0 otherwise). (7.21) ensures the condition that charging and discharging of the battery storage will not occur simultaneously. The storage level of the battery bank is defined by (7.6) and (7.22) ensures that the storage level is within the minimum and the maximum capacity of the storage bank. The operating time limits on the charging and discharging modes are defined by (7.23) and (7.24) respectively. Similarly, (7.25) and (7.26) define the motoring and generating mode power limits on the pump hydro storage unit. The m_{ph} and n_{ph} are the binary variables indicating the mode of operation of the pump hydro unit (m_{ph} = 1 means motoring mode, 0 otherwise, and if n_{ph} = 1 means generating mode, 0 otherwise). (7.27) ensures the condition that motoring and generating modes of the pump hydro unit will not occur simultaneously. The storage level of the pump hydro storage is defined by (7.8) and (7.28) ensures that the storage level is within the minimum and the maximum capacity of the pump hydro storage. The operating time limits on the motoring and generating modes are defined by (7.29) and (7.30) respectively. In addition to these constraints (7.31) represents the power limits (minimum and maximum) on the adjustable loads of the system, the operating time limit is defined in (7.32), and (7.33) describes the energy consumption limits on the loads over an operating cycle of 24 hours.

7.4 Simulation Results and Discussion

The system under investigation has three dispatchable diesel generators, one combined heat, and power plant, one boiler unit, two renewable plants solar and wind one each, one dual energy storage system comprising one battery and pump hydro, five adjustable loads, one fixed load and with limited power exchange capability. Table 7.2 and Table 7.3 represent the cost coefficients and other characteristics of the schedulable, and non-schedulable storage units, respectively. Table 7.4 shows the characteristics of adjustable loads of the microgrid system. Figure 7.1 and Figure 7.2 show the dynamic electrical and thermal load demands of the system and the energy price in the electricity market, respectively. Figure 7.3 shows the intermittent renewable power generation from solar and wind power plants along with the net aggregated renewable power generation of the microgrid. The three different operating cases have been modeled, simulated, and analyzed to obtain the impact of the minimum operating time on the net operating cost of the grid-connected microgrid system under study as given below:

- Case 1. Microgrid operating only with the battery storage system.
- Case 2. Microgrid operating only with the pump hydro storage system.
- Case 3. Microgrid operating with dual-energy storage (battery and pump hydro storage) system.

The 24-hour optimal scheduling has been obtained for a grid-connected microgrid system for three operating cases. The intermittency of the renewable solar and wind turbine has been considered in optimization modeling using the variations in power generation over the scheduling horizon. The maximum power exchange possible between the grid and the microgrid has been assumed to be 15 MW. The balance between the power supply and demand in the present system has been managed through limited power exchange possible between the grid and the microgrid along with controlling the minimum operating time of the storage system either battery, pump hydro storage or dual energy storage. The minimum operating

Table 7.2 Cost coefficients of the power generating units of MG.

Power plant		Fuel cost coefficient		
		a	b	c
Diesel Generator	Gen1	0.0005	21.63	1.054
	Gen 2	0.0005	21.63	1.054
	Gen 3	0.0025	9.87	1.054
CHP1	Gen4	0.22221	45.81	0.800
CHP2	Gen5	0.1000	51.60	0.461
Boiler		-	0.63	-
Pump hydro		-	0.06	-
Battery bank		-	1.06	-
Wind		-	-	-
Solar		-	-	-

Table 7.3 Characteristics of the power generating units of MG.

Power plants		Power generation			UP and Down rate of power generation		Minimum ON and OFF time		Storage level	
		$Pgen_{min}$	$Pgen_{max}$	Efficiency					P_{min}	P_{max}
Diesel Generator	Gen1	1	5	-	2.5	2.5	3	3	-	-
	Gen2	1	5	-	2.5	2.5	3	3	-	-
	Gen 3	0.80	3.00	-	3	3	1	1	-	-
CHP1		0.010	0.060	0.60	-	-	-	-	-	-
CHP2		0.010	0.060	0.60	-	-	-	-	-	-
Wind		0	1	-	-	-	-	-	-	-
Solar		0	1.5	-	-	-	-	-	-	-
Boiler		0	3	-	-	-	-	-	-	-
Pump hydro		0.4	2	0.90	-	-	3	2	4	20
Battery bank		0.4	2	0.90	-	-	3	2	4	20

Table 7.4 Data of shiftable and controllable loads.

Name of load	Type of load	Capacity		Energy consumed	Operating time		Min up time
		Min (MW)	Max (MW)	(MWh)	Initial (h)	Final (h)	(h)
Load 1	Shiftable	0.00	0.4	2.8	8	15	1
Load 2	Shiftable	0.00	0.4	2.4	13	19	1
Load 3	Shiftable	0.02	0.8	2.4	16	18	1
Load 4	Shiftable	0.02	0.8	4.0	14	24	1
Load 5	Controllable	1.70	2.0	47	1	24	24

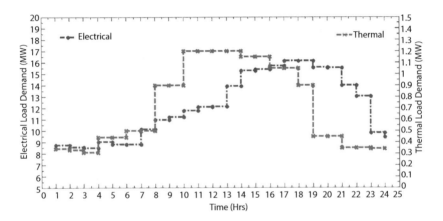

Figure 7.1 Electrical and thermal load demand of the microgrid system.

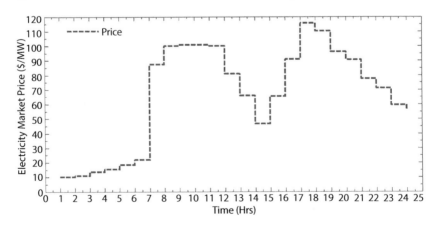

Figure 7.2 Day-ahead electricity market price.

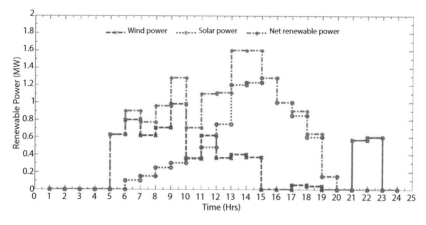

Figure 7.3 Solar, wind and net renewable power of the microgrid system.

time for charging/motoring and discharging/generating mode of respective storage system has been considered to be 2 Hrs and 3 Hrs, respectively. The impact of the minimum operating time on the economical operation of microgrid has been obtained and analyzed.

Figure 7.4 shows the variation in power exchange takes place between the grid and microgrid while operating in three operating case scenarios. It has been found that in case 3 microgrid operating with dual storage system requires less amount of power from the grid as compared to the microgrid system operating with one storage either battery (case 1) and pump hydro (case 2).

Microgrid systems can operate either with battery and pump hydro storage or with a dual storage system to manage the system variability. The operating

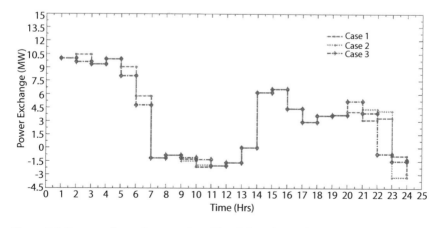

Figure 7.4 Power exchange between the grid and the microgrid.

conditions for pump hydro and the battery storage systems have been considered the same for the comparison purpose. The dual storage system counters the variability more effectively as compared to the individual storage system. The charging/motoring and discharging/generating schedules of both storage systems are depending upon the system power requirements and the state of charge at a particular instant of operation. In this study, the minimum operating time for charging/motoring and discharging/generating modes of battery/pump hydro storage system respectively have been assumed to be 2 hours and 3 hours. The state of charge has been considered to be same for the initial and final interval of scheduling. It is assumed to be 2MWh for the individual storage system and 1 MWh for each storage of dual energy storage system in current study. The operating schedule and the state of charge of the storage systems under all three operating cases have been shown in Figure 7.5 for cases 1 to 3. The variations in the state of charge of the storage systems indicate that the storage energy has been utilized to manage the power balance in the microgrid system through different modes of storage operations. Results show that the state of charge for case 1 and case 2 are identical except at the 21st and 22nd hours of an operation, mainly due to variations in the operating modes to minimize the operating cost. It has been also found that the net state of charge for dual storage systems is less as compared to microgrid system operation with one energy storage system either battery or pump hydro storage.

The variations in the power generation of the diesel generators have been shown in Figure 7.6(a)–Figure 7.6(c) for three cases of operation. The power generation from all three diesel generators varies according to the

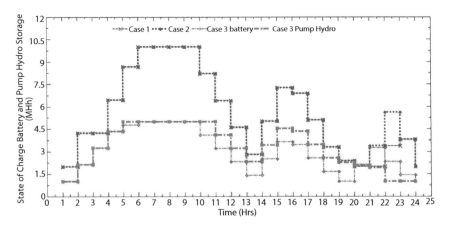

Figure 7.5 State of charge battery and pump hydro storage for single and dual storage operation.

Optimal Operation of GCMG with DESS 201

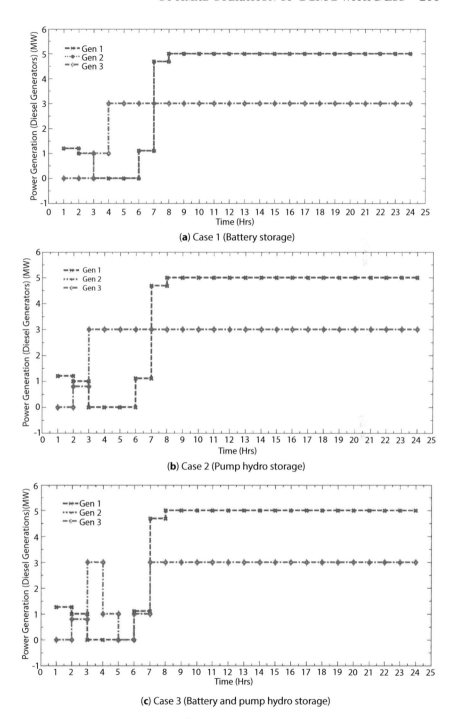

Figure 7.6 Power generation from diesel generators.

202 Energy Storage Technologies in Grid Modernization

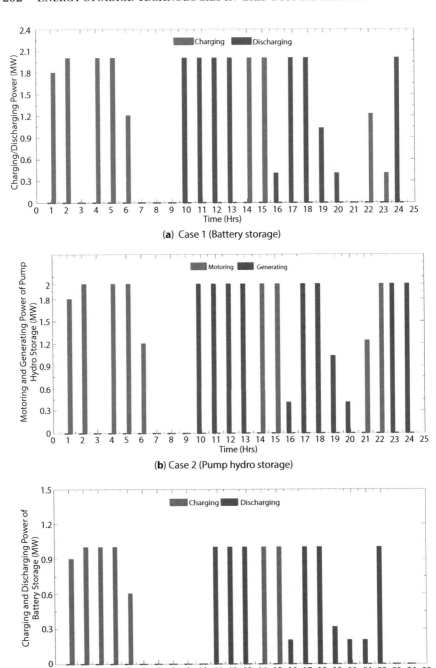

Figure 7.7 Charging/motoring and discharging/generating mode power of battery and pump hydro storage system. (*Continued*)

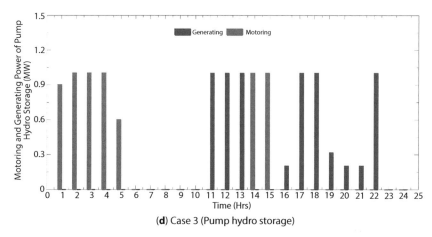

(**d**) Case 3 (Pump hydro storage)

Figure 7.7 (Continued) Charging/motoring and discharging/generating mode power of battery and pump hydro storage system.

price variations and state of storage. The power generation during the initial hours of scheduling is less, due to the higher state of charge available to meet the system demand. Figure 7.7(a)–Figure 7.7(b) represents the charging-discharging and motoring-generating modes of operation during the operational cases of microgrid when operated with battery storage and pump hydro storage individually, respectively. Figure 7.7(c)–Figure 7.7(d) show the variations of the charging-discharging and motoring-generating modes of operation separately during the microgrid operation with DESS. The pump hydro storage provides more charging power and therefore maintains a higher level of charge.

The microgrid system under study has five adjustable loads with operating time shifting and power controlling option during operating time depending upon the power availability and market energy prices. Figures 7.8(a)–(c) shows the scheduling of the adjustable load and it was found that the shiftable loads (load 1 – load 4) and controllable loads operate as per the operational and technical constraints to provide little flexibility to the system for managing optimal dispatch. Different scheduling has been noticed for different operating cases. The results obtained show that the loads operates within the time slots mentioned for the operations and consumed energy as per specifications given in Table 7.3. These variations in operating times of the various loads are due to the market prices and the power availability from other economical resources.

The operating cost of the GCMG system under study includes the cost of power generation from all energy resources, cost of charging/motoring and discharging/generating modes of operation of DESS, cost of power

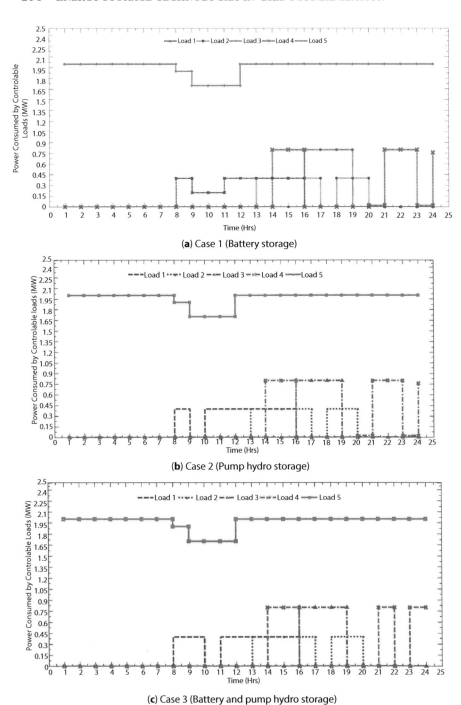

(a) Case 1 (Battery storage)

(b) Case 2 (Pump hydro storage)

(c) Case 3 (Battery and pump hydro storage)

Figure 7.8 Adjustable load scheduling.

Table 7.5 Costs associated with the scheduling of the microgrid operating under different cases.

Minimum operating time		Cost of power generation (diesel unit) ($)	Cost of power exchange ($)	Cost of power generation (CHP plant) ($)	Cost of power generation (boiler plant) ($)	Cost of battery power (storage) ($)	Cost of pump hydro power (storage) ($)	Total operating cost ($)
Charging/ motoring (hrs)	Discharging/ generating (hrs)							
			Case 1 (Battery storage only)					
1	1	4731.9	1706.6	170.56	771.12	38.006	*****	7418.2
1	2	4759.5	1648.7	170.56	771.12	35.306	*****	7385.2
1	3	4731.9	1720.4	170.56	771.12	33.174	*****	7427.1
2	1	4731.9	1719.4	170.56	771.12	34.169	*****	7427.1
2	2	4731.9	1743.1	170.56	771.12	32.274	*****	7449
2	3	4731.9	1743.1	170.56	771.12	32.274	*****	7449
3	1	4731.9	1558.4	170.56	771.12	33.174	*****	7265.1
3	2	4731.9	1556.4	170.56	771.12	33.174	*****	**7263.1**
3	3	4731.9	1763.2	170.56	771.12	33.174	*****	7469.9

(*Continued*)

Table 7.5 Costs associated with the scheduling of the microgrid operating under different cases. (*Continued*)

Minimum operating time		Cost of power generation (diesel unit) ($)	Cost of power exchange ($)	Cost of power generation (CHP plant) ($)	Cost of power generation (boiler plant) ($)	Cost of battery power (storage) ($)	Cost of pump hydro power (storage) ($)	Total operating cost ($)
Charging/ motoring (hrs)	Discharging/ generating (hrs)							
\multicolumn{9}{c}{Case 2 (Pump hydro storage only)}								
1	1	4759.5	1686	170.56	771.12	*****	2.162	7389.4
1	2	4731.9	1643.4	170.56	771.12	*****	1.9984	7319
1	3	4731.9	1720.4	170.56	771.12	*****	1.8778	7395.9
2	1	4694.7	1758.6	170.56	771.12	*****	2.191	7397.2
2	2	4731.9	1693	170.56	771.12	*****	2.044	7368.7
2	3	4731.9	1693	170.56	771.12	*****	2.044	7368.7
3	1	4731.9	1522.2	170.56	771.12	*****	2.095	7197.9
3	**2**	**4731.9**	**1520.2**	**170.56**	**771.12**	*****	**2.095**	**7195.8**
3	3	4759.5	1733.8	170.56	771.12	*****	1.8778	7436.9

(*Continued*)

Table 7.5 Costs associated with the scheduling of the microgrid operating under different cases. (Continued)

Minimum operating time		Cost of power generation (diesel unit) ($)	Cost of power exchange ($)	Cost of power generation (CHP plant) ($)	Cost of power generation (boiler plant) ($)	Cost of battery power (storage) ($)	Cost of pump hydro power (storage) ($)	Total operating cost ($)
Charging/ motoring (hrs)	Discharging/ generating (hrs)							
Case 3 (Dual energy storage)								
1	1	4692.8	1886.1	170.56	771.12	15.166	1.0142	7536.8
1	2	**4759.5**	**1755.2**	**170.56**	**771.12**	**14.692**	**0.9657**	**7472.1**
1	3	4731.9	1819.5	170.56	771.12	16.137	0.93889	7510.2
2	1	4731.9	1813.6	170.56	771.12	15.166	0.96704	7503.3
2	2	4731.89	1851.1	170.56	771.12	15.237	0.913	7540.86
2	3	**4759.5**	**1774.7**	**170.56**	**771.12**	**14.218**	**1.073**	**7491.3**
3	1	4759.54	1876.32	170.56	771.12	13.318	0.971	7591.84
3	2	4759.5	1864.1	170.56	771.12	13.318	0.93889	7579.6
3	3	4759.5	1791.1	170.56	771.12	15.237	0.93889	7508.5

exchange between the grid and the microgrid. In this present study, it is assumed that the microgrid purchase power from the grid at 80 percent and sell power to grid at 120 percent of the energy price rate at a particular time interval. The overall operational cost of the microgrid changes with the operating conditions. The operating cost of microgrid system associated with different operating cases have been tabulated in Table 7.5.

The minimum operating cost of $7,263.1 has been observed in case 1 when the minimum battery charging and discharging time fixed to 3 and 2 hours, respectively. This is due to the minimum power exchange cost for maintaining the power balance. Similarly, in case 2 the minimum operating cost of $7,195.8 has been observed for the minimum operating time limit of 3 and 2 hours, respectively. The operating cost in case of case 2 is less than in case 1 due to the less cost of operation of pump hydro and the power exchange as compared to the battery storage system. But during a dual energy storage operation (case 3), the cost has been found that system cost increases if the microgrid operates at the same operating time conditions on the storage. It has been found the optimal cost of operation of $7,472.1 when storage operates with a minimum time of operation limits of 1 hour of charging/motoring and 2 hours of discharging/generating mode. It has been concluded that the minimum time of operation of the storage system has a great impact on the optimal operation of the microgrid system and therefore the minimum time of operation limit must be considered carefully.

7.5 Conclusion

This chapter presents a grid-connected microgrid system having a dual energy storage system for optimal scheduling while operating with a minimum operating time constraint on the storage system. Day-ahead electricity market scheduling has been obtained for the microgrid system involving solar, wind, diesel, boiler, combined heat, and power plants and fixed and adjustable loads. Thermal load demand is considered to be fulfilled by the combined heat and power units and the boiler system. The microgrid manages the load demand very economically and effectively when the optimal operating minimum time of operation on the storage systems has been applied. The results obtained from the simulations confirm the impact of the minimum operating time limit set by the system operator on the storage systems on the operating cost the system. The dual storage systems are a little costlier than the individual storage system of same capacity but helps in reducing effectively net power demand variations of the system.

At the same time it reduces the net carbon emission by reducing the power generation from the diesel generators or generators using fossil fuels for generations. Therefore, it is recommended that the microgrid system with the flexible dual system with flexible operating time limits is more suitable for managing the system variability as compared to the system operating with battery storage and pump hydro storage systems individually.

References

1. V. V. S. N. Murty and A. Kumar, "Multi-objective energy management in microgrids with hybrid energy sources and battery energy storage systems," Prot. Control Mod. Power Syst., vol. 5, no. 2, pp. 1–20, 2020.
2. D. Kumar, Y. Verma, and R. Khanna, "Demand Response based Dynamic Dispatch of Microgrid System in Hybrid Electricity Market," Int. J. Energy Sect. Manag., vol. 13, no. 2, pp. 318–340, 2019.
3. M. Elsisi, M. Q. Tran, H. M. Hasanien, R. A. Turky, F. Albalawi, and S. S. M. Ghoneim, "Robust model predictive control paradigm for automatic voltage regulators against uncertainty based on optimization algorithms," Mathematics, vol. 9, no. 22, 2021.
4. M. Soshinskaya, W. H. J. Crijns-Graus, J. M. Guerrero, and J. C. Vasquez, "Microgrids: Experiences, barriers and success factors," Renew. Sustain. Energy Rev., vol. 40, pp. 659–672, 2014.
5. A. Rezaee Jordehi, M. S. Javadi, and J. P. S. Catalão, "Optimal placement of battery swap stations in microgrids with micro pumped hydro storage systems, photovoltaic, wind and geothermal distributed generators," Int. J. Electr. Power Energy Syst., vol. 125, p. 106483, 2021.
6. A. Majzoobi and A. Khodaei, "Application of microgrids in providing ancillary services to the utility grid," Energy, vol. 123, pp. 555–563, 2017.
7. A. Shahmohammadi, R. Sioshansi, A. J. Conejo, and S. Afsharnia, "The role of energy storage in mitigating ramping inefficiencies caused by variable renewable generation," Energy Convers. Manag., vol. 162, pp. 307–320, 2018.
8. M. Aneke and M. Wang, "Energy storage technologies and real life applications – A state of the art review," Appl. Energy, vol. 179, pp. 350–377, 2016.
9. D. Kumar, Y. P. Verma, R. Khanna, and P. Gupta, "Impact of market prices on energy scheduling of microgrid operating with renewable energy sources and storage," Mater. Today Proc., vol. 28, pp. 1649–1655, 2020.
10. L. HL, Z. ZQ, T. XJ, C. Zheng, S. Li, and J. Yang, "Research on optimal capacity of large wind power considering joint operation with pumped hydro storage," Power Syst Technol, vol. 39, no. 10, pp. 2746–2750, 2015.
11. Y. Dvorkin, D. S. Kirschen, and M. A. Ortega-vazquez, "Assessing flexibility requirements in power systems," IET Gener. Transm. Distrib., vol. 8, no. 11, pp. 1820–1830, 2014.

12. A. J. Lamadrid and T. Mount, "Ancillary services in systems with high penetrations of renewable energy sources, the case of ramping," Energy Econ., vol. 34, no. 6, pp. 1959–1971, 2012.
13. G. Liu, Y. Xu, and K. Tomsovic, "Bidding Strategy for Microgrid in Day-Ahead Market Based on Hybrid Stochastic / Robust Optimization," IEEE Trans. Smart Grid, vol. 7, no. 1, pp. 227–237, 2016.
14. Y. P. Verma and A. Kumar, "Economic-emission unit commitment solution for wind integrated hybrid system," Int. J. Energy Sect. Manag., vol. 5, no. 2, pp. 287–305, 2011.
15. A. Chaouachi, R. M. Kamel, R. Andoulsi, and K. Nagasaka, "Multiobjective Intelligent Energy Management for a Microgrid," IEEE Trans. Ind. Electron., vol. 60, no. 4, pp. 1688–1699, 2013.
16. C. Chen, S. Duan, T. Cai, B. Liu, and G. Hu, "Smart energy management system for optimal microgrid economic operation," IET Renew. Power Gener., vol. 5, no. 3, pp. 258–267, 2011.
17. H. Shayeghi and B. Sobhani, "Integrated offering strategy for profit enhancement of distributed resources and demand response in microgrids considering system uncertainties," Energy Convers. Manag., vol. 87, pp. 765–777, 2014.
18. G. Ferruzzi, G. Cervone, L. Delle, G. Graditi, and F. Jacobone, "Optimal bidding in a Day-Ahead energy market for Micro Grid under uncertainty in renewable energy production," Energy, vol. 106, pp. 194–202, 2016.
19. P. Fazlalipour, M. Ehsan, and B. Mohammadi-Ivatloo, "Risk-aware stochastic bidding strategy of renewable micro-grids in day-ahead and real-time markets," Energy, vol. 171, pp. 689–700, 2019.
20. M. N. Faqiry and S. Das, "Double Auction with Hidden User Information: Application to Energy Transaction in Microgrid," IEEE Trans. Syst. Man, Cybern. Syst., vol. 49, no. 11, pp. 2326–2339, 2019.
21. D. Kumar, S. Dhundhara, Y. P. Verma, and R. Khanna, "Impact of optimal sized pump storage unit on microgrid operating cost and bidding in electricity market," J. Energy Storage, vol. 51, p. 104373, 2022.
22. M. B. Sigalo, A. C. Pillai, S. Das, and M. Abusara, "An energy management system for the control of battery storage in a grid-connected microgrid using mixed integer linear programming," Energies, vol. 14, no. 19, p. 6212, 2021.
23. A. F. Nematollahi, H. Shahinzadeh, H. Nafisi, B. Vahidi, Y. Amirat, and M. Benbouzid, "Sizing and sitting of DERs in active distribution networks incorporating load prevailing uncertainties using probabilistic approaches," Appl. Sci., vol. 11, no. 9, p. 4156, 2021.
24. N. Nasiri, S. Zeynali, S. N. Ravadanegh, and M. Marzband, "A hybrid robust-stochastic approach for strategic scheduling of a multi-energy system as a price-maker player in day-ahead wholesale market," Energy, vol. 235, p. 121398, 2021.

25. U. T. Salman, F. S. Al-Ismail, and M. Khalid, "Optimal sizing of battery energy storage for grid-connected and isolated wind-penetrated microgrid," IEEE Access, vol. 8, pp. 91129–91138, 2020.
26. D. Kumar, Y. P. Verma, R. Khanna, and P. Gupta, "Impact of market prices on energy scheduling of microgrid operating with renewable energy sources and storage," Mater. Today Proc., vol. 28, pp. 1649–1655, 2020.
27. D. Kumar, Y. P. Verma, and R. Khanna, "Consumer Participation Based Scheduling of Microgrid System in Electricity Market," in 2018 IEEE 8th Power India International Conference (PIICON), 2018, pp. 1–5.
28. L. Bagherzadeh, H. Shahinzadeh, H. Shayeghi, and G. B. Gharehpetian, "A short-term energy management of microgrids considering renewable energy resources, micro-compressed air energy storage and DRPs," Int. J. Renew. Energy Res., vol. 9, no. 4, pp. 1712–1723, 2019.
29. A. Ghasemi, "Coordination of pumped-storage unit and irrigation system with intermittent wind generation for intelligent energy management of an agricultural microgrid," Energy, vol. 142, pp. 1–13, 2018.
30. Y. Guo and C. Zhao, "Islanding-aware robust energy management for microgrids," IEEE Trans. Smart Grid, vol. 9, no. 2, pp. 1301–1309, 2018.
31. J. Moradi, H. Shahinzadeh, and A. Khandan, "A cooperative dispatch model for the coordination of the wind and pumped-storage generating companies in the day-ahead electricity market," Int. J. Renew. Energy Res., vol. 7, no. 4, pp. 2057–2067, 2017.
32. S. Sukumar, H. Mokhlis, S. Mekhilef, K. Naidu, and M. Karimi, "Mix-mode energy management strategy and battery sizing for economic operation of grid-tied microgrid," Energy, vol. 118, pp. 1322–1333, 2017.
33. Y. Xiang, J. Liu, and Y. Liu, "Robust Energy Management of Microgrid with Uncertain Renewable Generation and Load," IEEE Trans. Smart Grid, vol. 7, no. 2, pp. 1034–1043, 2016.
34. S. Das and M. Basu, "Day-ahead optimal bidding strategy of microgrid with demand response program considering uncertainties and outages of renewable energy resources," Energy, vol. 190, p. 116441, 2020.
35. S. Dhundhara and Y. P. Verma, "Application of micro pump hydro energy storage for reliable operation of microgrid system," IET Renew. Power Gener., vol. 14, no. 8, pp. 1368–1378, 2020.
36. S. Dhundhara, Y. P. Verma, and A. Williams, "Techno-economic analysis of the lithium-ion and lead-acid battery in microgrid systems," Energy Convers. Manag., vol. 177, pp. 122–142, 2018.
37. N. I. Nwulu and X. Xia, "Optimal dispatch for a microgrid incorporating renewables and demand response," Renew. Energy, vol. 101, pp. 16–28, 2017.
38. D. Kumar, Y. P. Verma, and R. Khanna, "Impact of Battery Scheduling on the Operation of Microgrid System having Dynamic Load Demand," 2019 Glob. Conf. Adv. Technol. GCAT 2019, Oct. 2019.

39. A. Khodaei, "Microgrid Optimal Scheduling With Multi-period Islanding Constraints," IEEE Trans. Power Syst., vol. 29, no. 3, pp. 1383–1392, 2014.

8

Applications of Energy Storage in Modern Power System through Demand-Side Management

Preeti Gupta* and Yajvender Pal Verma

Department of Electrical & Electronics Engineering, UIET, Panjab University, Chandigarh, India

Abstract

The growing need to reduce greenhouse gas emissions, and the emergence of a competitive electricity market all over the world, calls for optimal deployment of the existing generation facilities coupled with limiting the dependence on fossil fuels for electricity generation. Renewable energy resources provide a greener option for electricity generation. However, owing to their intermittency and stochasticity, the grid integration of these resources poses serious operational issues. The increased share of renewables in the generation mix augments net uncertainty and makes it more challenging to maintain supply-demand balance. One of the solutions to mitigate the variability of renewables is incorporating energy storage means. However, installing energy storage devices of suitable size is often associated with large capital costs. In this regard, Demand-Side Management (DSM) options can provide an efficient and cost-effective solution. DSM is the concept that focuses on enhancing the efficiency of various energy processes and also provides options for better utilizing the available generating facilities by adjusting load demand. This chapter provides a detailed introduction to the DSM. The detailed modelling of battery energy storage devices, DR models for minimizing the net load variations; and minimizing the peak load and peak to average ratio (PAR) with DR participation of residential Battery Energy Storage System (BESS) in the solar photo-voltaic integrated system is also discussed.

Keywords: DSM, demand response, PAR, renewable energy resources, BESS

---

*Corresponding author: preetigupta1999@gmail.com

Sandeep Dhundhara, Yajvender Pal Verma, and Ashwani Kumar (eds.) *Energy Storage Technologies in Grid Modernization*, (213–238) © 2023 Scrivener Publishing LLC

8.1 Introduction to Demand-Side Management

The development of new power system modules and the setting up of the competitive electricity market have driven the utilities to deliver more reliable, stable, and cheaper power to the end users. The advancements in information and communication technologies have added observability to the system and have enabled utilities to exercise control on the demand side. It has steered to the idea of Demand-Side Management (DSM). DSM can be described as the planning, implementation, and monitoring of electricity consumption activity designed to encourage consumers to amend their electricity usage pattern, both concerning timing and amount of electricity consumed so as to improve the overall efficiency and/or economy of the system [1]. DSM offers various benefits such as mitigating emergencies, reducing blackouts, and enhancing system reliability. The appropriate deployment of the DSM techniques may also reduce the energy reserve requirements and can help in reducing energy costs and harmful emissions. Various DSM techniques and their broad classifications are described in next section.

8.1.1 Demand-Side Management Techniques

The power management and operation of electrical energy systems have always been optimized from the generation and distribution perspective. The demand side has received attention only in the recent years leading to the evolution and implementation of the concept of DSM. This includes a variety of different techniques to enhance the overall system efficiency; and optimized utilization of infrastructural facilities and electrical energy. A broad classification of various DSM techniques has been shown in Figure 8.1 [2, 3].

8.1.1.1 Energy Efficiency

Energy efficiency is a long-term conservation strategy involving technological measures that save energy while maintaining the same or superior levels of energy services. The implementation of energy efficiency programs helps in reducing the peak demand, postponing the power system capacity enhancement, and reducing power system costs. All the initiatives to educate and motivate consumers to use more efficient devices, appliances and processes fall under this category [2, 4]. Following are some of the examples of energy-efficient approaches being adopted by the utilities:

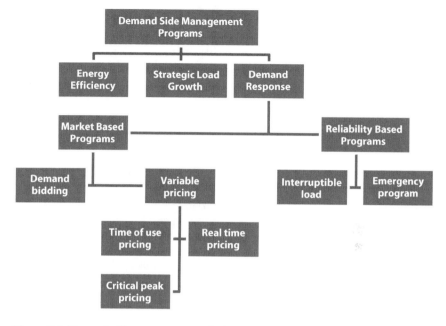

Figure 8.1 Demand-side management techniques.

1. Adopting energy-efficient appliances and replacing lighting with LED lights to reduce energy consumption and create awareness among the users about energy efficiency.
2. Encouraging regular maintenance of various electrical equipments, recovering waste heat, and implementing cogeneration.
3. Designing energy-efficient buildings for consistent reduced energy consumption and installing more efficient methods for space conditioning.

8.1.1.2 Demand Response

Demand response (DR) refers to short-term load management programs designed to influence energy consumption patterns. It is defined as "the ability of the end users to change the electricity consumption at their location from the normal consumption pattern in response to changes in electricity price, or to incentive payments aimed at inducing lesser electricity consumption at the high wholesale price points or when the system reliability is threatened." Given that DR provides the capability to the system operator to influence and manipulate the load directly, the DR programs are increasingly gaining importance in the new electricity market environment. The DR programs are broadly classified into two categories; market-based or price-based programs and reliability-based or incentive-based programs. In the market-based DR programs, customers have the option to adjust their consumption according to the real-time electricity price. The reliability-based DR programs offer incentives to the enrolled customers for voluntary and/or involuntary adjustment of their controlled appliances [5].

A brief description of the different market-based DR programs is given below:

I. *Demand bidding program*
   This program allows large consumers to submit their demand bids for specific load curtailment at a price lower than the market rate. On the acceptance of the bid in the wholesale market, it becomes mandatory for the consumers to curtail the specified bid amount. The deviations from the scheduled bidding are liable to be penalized.

II. *Real-time pricing programs*
   As the electricity industry moves towards a more modern grid - and more aware and educated end users; the utilities are beginning to introduce pricing structures that align the electricity tariff with its actual production cost by varying the electricity prices as per the time of its consumption. This is so because the electricity production cost is not the same, but fluctuates over time. As the peaking plants are often more expensive than the base load plants, the electricity costs are more during peak hours. These programs offer dynamic pricing structures aimed to limit peak time demands. Various real-time pricing program introduced by the utilities are discussed below:

   i) Time-of-use (ToU) rate: ToU rate is basically the segregation of energy rates based upon the time of energy consumption. This has been used for decades by utilities with large commercial and industrial consumers. ToU electricity prices are offered over a wide range of time scales, seasonal, monthly, weekly or daily. The varying energy rates reflect the basic electricity production costs during different time periods.

   ii) Critical peak pricing (CPP): CPP is one of the time-varying pricing plans in which the electricity is priced higher during certain peak demand periods.

Utilities often send prior notification (day ahead) of the CPP period so that consumers can be pre-prepared to reduce their electricity consumption. CPP can also be used to improve the system reliability as it reflects the system state. Thus, by sending appropriate CPP signals, the consumers can be motivated to decline the load during the system stress events.

iii) Real-time rate: Under this program, customers pay a tariff that is a function of the actual electricity market price. The rates are usually supplied hourly and vary in accordance with the fluctuations in the power supply. The time granularity may vary from system to system [6].

A brief description of different reliability-based DR programs is given below:

i) Interruptible load program: In this DR program, consumers are offered discounted electricity rates or incentives as compensation for allowing service interruptions. Interruptible load programs had been applied by the large commercial and industrial consumers earlier also. Nowadays, such types of programs are also being designed for small consumers as well.

ii) Direct load control program: In the direct load control DR programs, the utility is allowed to directly reduce or interrupt the electricity supply during peak hours or when the prices are high. The consumers receive compensation for the supply interruptions.

iii) Emergency program: These programs are designed to manage the system load demand during system contingencies. The consumers are compensated by receiving incentives for declining their demand during system emergency conditions. In contrast to the other two programs, penalties may also be imposed if they abate from the program when required.

8.1.2 Demand-Side Management Approaches

Various DSM approaches have been used by the utilities to achieve the desired load curves [3, 7]. The particular techniques are presented in Table 8.1.

Table 8.1 Different DSM approaches and their objectives.

Type	Objectives
Peak shaving or peak clipping	Peak clipping approach reduces the load during peak periods without compensating the same at some other period.
Valley filling	Deferring demand to off-peak periods; thus reducing the difference between off-peak and peak periods. Hence improving the network load factor.
Load shifting	Flattening of load curve by shifting load from peak periods to off-peak period resulting in load reduction at one period and increase by the same amount in some other time period.
Strategic Energy conservation	The main objective is to reduce losses in the system at the user end. Thus improving terminal system efficiency. Correction of lighting systems is an example of strategic energy saving.
Strategic load growth	Increasing the general load demand over the consumption profile.
Flexible load shape	It involves the identification and control of the customers who are willing to be controlled during emergency periods through the smart grid management system.

8.2 Operational Aspects of DR

Demand response is an important tool that enables end-user participation in handling different operational issues. It helps energy providers to reduce the peak demand and also align the energy consumption with the non-dispatchable renewable generation. On the consumer side, DR allows them to select the energy provider of their choice, earn incentives for DR participation or reduce electricity bills by changing their energy consumption pattern. This helps in reducing the overall operational cost and carbon emission levels as well as increasing the grid sustainability. Some of the operational aspects of DR as highlighted in Figure 8.2 are discussed below [8, 9].

i) Load profile management
Demand response is not a new concept but in recent years it has been receiving more attention. Due to the developments in information and

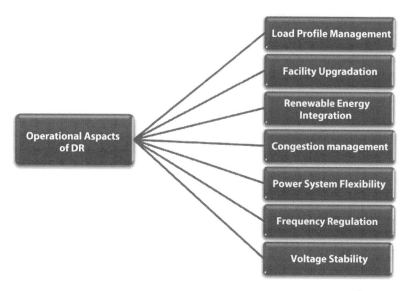

Figure 8.2 DR operational aspects.

wireless communication technologies and wide applications in the power systems like smart meters and energy management systems, participation of small consumers in DR has also been made possible.

Peak-load management is an important function, wherein utilities reshape the load curve to reduce overall operational cost and improve energy efficiency. Utilizing load-shifting DR programs with a large number of devices of various types, the power consumption can be shifted from the peak load periods to more appropriate times so that the distribution system's peak demand can be reduced [10].

ii) Facility upgradation

Application of DR may effectively delay the requirement for transmission and distribution system upgradation. DR techniques can be applied to adjust the power consumption from heavily loaded periods to light load periods and ESS can be used to reduce the impact of peak load. DR aims to match the power demand with the available resources without adding new generation capacity by affecting peak shaving and valley filling. Therefore, the need for expensive capacity enhancement can be postponed [11, 12].

iii) Renewable energy integration

Over the last two decades, the utilization of renewable energy sources, and in particular large wind and solar energy into electricity generation, is on the rise. The displacement of conventional energy resources by renewable energy is considered to be a sustainable option for reducing greenhouse gas emissions. However, compared with fossil fuel–based electricity generation, wind and solar have challenging non-dispatchable operating characteristics such as variable and lower capacity factors and intermittent, variable availability [13].

DR is believed to provide one of the efficient solutions to accommodate the increasing penetration of the variable renewables with uncertainties and volatilities. The controllable loads and energy storage can be operated to align the system load demand with the generation. The power consumption can be reduced by exerting proper DR when lesser electricity is being produced by the renewable generators [14].

iv) Congestion management

Proper application of DR can be effective in handling congestion in the distribution systems. Typically, consumers are accustomed to switching on appliances randomly and consuming electricity without interruption. With the bi-directional communication between the consumers and system operator becoming reality, some of the residential loads can be treated as flexible and controllable. The time of operation of such load can be adjusted in response to the system operating conditions. Thus, the flexible loads and ESS provide the system operator with the capacity to ease the network congestion points. The appliances that are time–flexible could be used to ease the network congestion points. This can include airconditioners, refrigerators, and water heaters [15, 16].

v) Power system flexibility

DR is a fast-growing resource management strategy that yields a dual impact of managing electricity demand and enabling efficient and flexible system operation. Utilities with a larger share of variable renewable generation find it more difficult to follow the load curve and meet the power demand than the systems with conventional generators only. The conventional generators are capable of providing spinning reserve capacity to adjust the load variations and maintain the vital supply-demand balance in the power grid. However, in highly renewable energy mixed systems, it becomes difficult for the system operator to manage the required spinning reserve capacity efficiently. DR can provide alternative means to create indirect reserve capacity and implement cost-efficient flexibility [17].

vi) Frequency regulation

System frequency is a direct measure of the balance between system demand and generation at any point in time and must always be maintained within narrow statutory bounds to ensure system stability. The inherent stochasticity of the load demand makes it difficult for the system operators to maintain this balance. Sufficient spinning reserves and fast-responding generators are required to meet the load uncertainties. In addition, sufficient generation capacity is required to meet the peak load demand, which may occur only for small periods during the day. This adds to the cost of the electricity and inefficient use of the generation capacity. Hence, the generation utilities used different tariff schemes to encourage more electricity utilization during off-peak hours and vice versa. Earlier, only large industrial consumers used to take part in this activity. More recently, with the advancements in technology, there have been initiatives to incorporate electrical appliances also into the demand-side management activity and provide frequency regulation services. The time flexible electrical appliances including residential, commercial, or industrial water heaters, air conditioners, and refrigerators could be used for this purpose [18].

vii) Voltage stability
The voltage level in power transmission and distribution networks in the grid must be maintained within a specified tolerance band. The power distribution systems are persistently facing ever-growing load demand. Distribution networks experience wide load variations every day, distinctly varying from low to high levels. The large variations in load if not managed properly may result in voltage fluctuations. In certain heavily loaded areas, under critical loading conditions, the distribution network may experience voltage collapse [19].

One of the promising solutions to enhance the voltage stability in the distribution networks is through end-user participation in DR programs. DR participation of ESS may contribute to maintain required voltage stability margin at network nodes by injecting and absorbing the controllable reactive power [20].

8.3 DSM Challenges

The concept of demand-side management is not new. However, widespread implementation of DSM is lacking. Although the key technologies required for its implementation have been developed, several challenges need to be addressed to increase DR participation [8, 21]. These include:

i) Lack of infrastructure
There has been a substantial development in communications and information technologies in recent years. But the deployment of advanced

information, communications, and metering and control technologies in most of the electricity systems has not been there. The integration of various demand response techniques in the system operations requires the deployment of appropriate sensors and advanced measurement and control devices in addition to sophisticated metering and trading facilities. In order to support the real-time execution of various DR programs a wide range of information, communication, and control systems are needed and control actions involving generation, loads and other network devices are required [1].

ii) Market barriers
The implementation of enabling technologies like DSM often involves different participants. The market penetration of DR programs and storage is influenced and hampered by factors like market structure, financing, incentives, institutional, policy, and regulatory framework. Considering the characterization and disaggregation of their multi-stream value, the development of a business case for such technologies becomes very complex. In the existing deregulatory market structure, different sectors (e.g., generating companies, transmission and distribution companies, etc.) of the power industry are operating as individual businesses. The benefits of DR participation can be associated with several power industry sectors that may be willing to deploy and incentivise certain specific aspects of the activity. There are multiple recipients of the DR activity who are willing to use it for their individual gains. Clearly, the individual power sector will tend to utilize the services for its own interests without caring for maximising the overall system benefits that could only be achieved by trading off the individual segment benefits. Thus, a suitable regulatory framework is a must to extract the benefits of DR and storage technologies in the current deregulated system [4].

Another important aspect is that the increased DR activity at the distribution level should not hamper the secure and safe operation of the power grid. In this context, a holistic approach and a clear regulatory mechanism are essential to ensure that no additional voltage violations and congestion are introduced by DR.

iii) Increased system complexity
Another barrier to the implementation of the DR is the increased operational complexity. The integration of DR in the system operation tends to increase the system complexity due to added information and control requirements. Nevertheless, given the flexibility requirements with the increased penetration of the renewable-based generation; to deal with the uncertainties DR is now been seen as an important tool in future

developments. Moreover, as the number of utilities implementing DR is increasing coupled with developments in technology, there is a continuous fall in the DR implementation cost. It can be expected in near future that the technology will become more competitive and gain improved confidence in the application of DR schemes.

iv) Behavioural barriers
Implementation of the DR schemes involves the participation of different categories of consumers. The DR participation of small consumers can be influenced not only by the economic benefits. Rather it may be more driven by other factors like convenience, comfort trust, and credibility. The trust and credibility of the DR provider play a significant role in the acceptance of DR programs, especially in the case of smaller, local DR participants whom they identify with trust. In particular, it can be linked to the resistance to allow third-party information and control of appliance usage patterns.

v) Safety and security of data
The safety and security of the data collected under the DR actions are very important and crucial. At the data flow level, it is imperative to have a secure, safe, secure, and efficient transfer and sharing of data with the different entities involved in the DR action. This is a very important and vital factor as it plays a significant role in building the DR provider credibility and helps in the long-term association between DR participants and the DR providers and ultimately effective implementation of the DR program.

8.4 Demand Response Resources

Different options that could be available for DR participation have been shown in Figure 8.3. DG resources including wind and solar photovoltaic generation are intermittent and generally uncontrollable whereas DGs such as diesel or gas generators or small Combined Heat and Power Plants (CHPs) are controllable [22, 23]. The distributed energy resources are capable of reducing the net load demand of the utility effectively during peak hours. However, vice versa is not true in their case, as these cannot be utilized for increasing the demand during low load periods.

Interruptible loads include the portion of load demand that is not very critical. This is generally comprised of a set of those devices whose operation is not very important and their disconnection does not cause much inconvenience to the consumers. The DR participation of interruptible loads also helps in reducing the demand during peak hours only. Deferrable

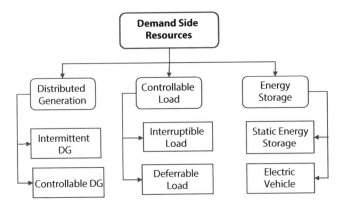

Figure 8.3 Different DR resources.

loads include appliances that are time flexible in operation. The deferrable devices DR can shift energy consumption rather than curtail within their operational bounds. These can be thought of as storage and utilized both for absorption as well as the release of energy around their nominal operational and power consumption trajectory.

Another DR resource is offered by the energy storage devices available on the consumer side. These may include both static energy storage systems (such as batteries and electrical heating systems) and Electric Vehicles (EVs). The choice of energy storage for DR is motivated by the fact that they can absorb and supply controllable energy on predetermined charge and discharge rates. A considerable amount of balancing capability and flexibility can be provided by a collection of small consumer-side energy storage devices. These could be an effective DR source, especially in variable renewable integrated systems by offering a large collective energy storage capacity. In the next section, the role of battery energy storage DR in solar PV integrated system has been explained in detail.

8.5 Role of Battery Energy Storage in DSM

Battery energy storage is a controllable and dispatchable resource and can be utilized effectively in response to power system situations such as price fluctuation or renewable energy intermittency [24]. Storage batteries have a fast response time and can be deployed to overcome the variability of load and renewable generation. However, if the energy storage devices are installed to solely handle the intermittency of renewable generation, the investment cost will increase substantially as the storage capacities required will be very large.

To reduce the large-scale deployment of energy storage at the supply-side, consumer-side energy storage devices can be incorporated through suitably designed demand response programs. In this section, applications of residential consumers' battery energy storage systems DR have been explained through two separate case studies. In the first case study, battery energy storage DR has been applied to reduce the PAR, and hence peak demand and valley filling have been achieved in a PV integrated system. In the second case study, the net load profile variations from the forecasted load have been minimized using the storage capacity of the residential BESS through DR.

8.5.1 Case Study I: Peak Load and PAR Reduction

In a PV integrated distribution system, the availability of solar generation reduces the net grid power demand during daytime; however, it has been observed that the demand curve usually peaks during evenings. Thus, connecting PV generation does not help in reducing the system peak. This case study explains the application of consumer-side batteries to store the energy during the low net load periods and dispatching during the high demand period when PV power is not available. This helps in reducing the peak demand and improving the peak-to-average ratio. The actual load curve of Chandigarh, a union territory city of India, is used for the study. The city has a total of 1,87,687 residential electricity consumers. The study has been performed for a set of 1,000 consumers, thus a modified load curve is obtained by scaling down the actual load profile. It has been assumed that 70% of the consumers participate in DR.

8.5.1.1 Problem Formulation

The objective function can be defined as (8.1):

$$Min \sum_{t \in N} abs(Pav_{dev}^t) \tag{8.1}$$

Where, N is the total number of the time slot in 15-minute blocks and Pav_{dev}^t is the deviation of the net modified demand from the average demand during t^{th} time slot. Net demand P_{net}^t is the remaining demand after reducing the solar generation, P_{pv}^t from the actual load demand P_{ac}^t. The net demand curve is modified through power exchange with the batteries and the values of modified load demand at any time $P_{net,mod}^t$ can be obtained as in (8.2) and (8.3) below:

$$P_{net}^t = P_{ac}^t - P_{pv}^t \qquad (8.2)$$

$$P_{net,mod}^t = P_{net}^t - P_{bat}^t \qquad (8.3)$$

Here, P_{bat}^t is the cumulative power exchange with the batteries of all the households employed in DR. The energy storage capacity of the participating batteries is conducted by the network operator to shave the peak demand of the net load by reducing the gap between the actual net demand and the average net demand Pav_{dev}^t by eq. (8.4).

$$Pav_{dev}^t = Pav_{net} - P_{net,mod}^t \qquad (8.4)$$

Here, Pav_{net} is the average value of the net demand over the settlement period. The energy storage capacity of the batteries is utilized to shift the load from the peak hours to the minimum net load periods. The central operator coordinates the charging and discharging cycles of the batteries to reduce peak demand and fill the valley area of the net demand curve.

8.5.1.2 Energy Storage Dispatch Modelling

The battery energy storage DR strategy implemented here works on the principle of controlling the charging and discharging of the consumer-owned batteries so as to minimize the gap between the average load and the actual load during each time block. The energy storage capacity is operated by the distribution network operator (DNO) to shave peak demand. The specifications of the batteries of the assumed BESS are given in Table 8.2. The mathematical formulation of the battery power exchange is given below in (8.5):

$$P_{bat}^t = P_{bat,dis}^t \times T_{dis}^t - P_{bat,ch}^t \times T_{ch}^t \qquad (8.5)$$

Here $P_{bat,dis}^t$ and $P_{bat,ch}^t$ are the charging and discharging power dispatched by the batteries and are represented as given in (8.6) and (8.7). To mark a time block as either discharging or charging, two 96X1 column vectors T_{dis}^t and T_{ch}^t are defined. The elements in these two vectors are either 0 or 1. In the vector T_{dis}^t, 1 represents discharging operation in the time block while 0 is not, and on the contrary in the vector T_{ch}^t, 1 represents charging operation while 0 is not.

$$P_{bat,dis}^t = \frac{n_{dis}^t \times p_{bat}}{\eta_{dis}} \tag{8.6}$$

$$P_{bat,ch}^t = n_{ch}^t \times p_{bat} \times \eta_{ch} \tag{8.7}$$

Here, n_{dis}^t and n_{ch}^t are the number of batteries undergoing discharging and charging cycles respectively. The efficiency of batteries is not the same during charging and discharging cycles and these have been represented by η_{ch}, η_{dis} respectively. The power exchange with the battery can be obtained from terminal voltage and the current flowing, and is limited by the maximum allowable battery current. Assuming that battery operation at maximum allowable current and constant terminal voltage, the battery power exchange during any operating cycle p_{bat} is constant.

The charging/discharging operation during any time block changes the state of charge (SoC) of the batteries. During charging cycles the energy is stored in the batteries and hence, the SoC of the batteries improves while during discharging cycles the state of charge of batteries deteriorates as the stored energy is dispatched. Equations relating $P_{bat,dis}^t$ and $P_{bat,ch}^t$ with the SoC are given as in (8.8), (8.9):

$$SoC^t - SoC^{t-1} = \Delta t \times P_{bat,ch}^t \tag{8.8}$$

$$SoC^{t-1} - SoC^t = \Delta t \times P_{bat,dis}^t \tag{8.9}$$

The operation of the battery storage system must satisfy the following constraints:

- The charging/discharging power during a time block is limited by the maximum allowable energy transfer through the batteries. This is determined by the battery terminal voltage and the maximum allowable battery current. By assuming the terminal voltage of batteries to remain constant at their rated value and keeping current at its maximum permissible value P_{bat}^{max} can be obtained. This constraint is formulated as (8.10), (8.11):

$$0 \leq P_{bat,ch}^{t} \leq P_{bat}^{max} \quad (8.10)$$

$$0 \leq P_{bat,dis}^{t} \leq P_{bat}^{max} \quad (8.11)$$

- The state of charge of the battery is limited by maximum energy storage capacity SoC_{max} and minimum permissible charging level SoC_{min} which depends upon the battery depth of discharge specification. Thus, battery SoC during a time block is subject to the following constraint (8.12):

$$SoC_{min} \leq SoC^{t} \leq SoC_{max} \quad (8.12)$$

- The number of batteries charging/discharging during a time block cannot be more than the number of residents participating in DR (n_{dr}). This is given as (8.13):

$$n_{dis}^{t}, n_{ch}^{t} \leq n_{dr} \quad (8.13)$$

In addition to the above constraints, it has been assumed that the battery SoC at the end of the settlement period is equal to the SoC at the beginning of the settlement period. Thus, the total load supplied or the energy drawn over the day with and without DR is the same. This can be formulated as (8.14):

Table 8.2 Battery specifications [25].

Description	Unit
Battery Capacity	2.4kWh
Depth of discharge limited	2kWh (80%)
Charging/discharging current limit	<20% of rated AH
Round trip efficiency	80%

$$\sum_t P_{net}^t = \sum_t P_{net,\text{mod}}^t \qquad (8.14)$$

8.5.2 Case Study II: Minimizing Load Profile Variations

Load curve fluctuations and uncertainty are very common and critical issues in power system operations. They have been the main cause of load–supply imbalance and power system stability problems [26]. Integration of variable renewables on the load side further increases the uncertainty and the stochasticity of the net load curve (the remaining demand that is not supplied by RES). The distribution utilities are required to purchase more reserves for supplying higher uncertainties, which increases the operational costs significantly. In this regard, DR options are now being looked upon to provide an efficient and cost-effective solution [27].

In this case study, the battery energy storage DR has been applied to minimize the disparity between the predicted net load profile and the actual net load profile. In this case, also, centralized DR participation of the residential community through an incentive-based DR scheme has been considered. The case of a distribution utility with residential, commercial, and industrial consumers has been assumed for investigation. It has been assumed that there are 1,000 residential consumers and 70% of them enrol for energy storage DR. The specifications assumed for the batteries are given in Table 8.2.

8.5.2.1 Problem Formulation

The objective function is defined by (8.15) as follows:

$$\text{Min} \sum_t abs\left(P_{sch}^{t,net} - P_{ac}^{t,net}\right) \qquad (8.15)$$

Here, $P_{sch}^{t,net}$, and $P_{ac}^{t,net}$ are the net forecasted and net actual load profiles respectively. Net scheduled load is obtained by reducing the forecasted SPV generation $P_{cal}^{t,pv}$ from the forecasted load P_{sch}^t. The forecasted SPV generation is calculated using forecasted solar irradiance. The net actual load profile is obtained by reducing the actual SPV generation $P_{ac}^{t,pv}$ from the actual load P_{ac}^t during a time slot (8.16), (8.17).

$$P_{sch}^{t,net} = P_{sch}^{t} - P_{cal}^{t,pv} \tag{8.16}$$

$$P_{ac}^{t,net} = P_{ac}^{t} - P_{ac}^{t,pv} \tag{8.17}$$

Using smart energy management systems and smart metering devices the charging/discharging operation of the BESS within the consumers' premises can be implemented [28]. The deviations of the actual net load profile from the forecasted are minimized by controlling battery power exchange. The power balance equation considering BESS DR can be written as (8.18):

$$P_{sch}^{t,net} - P_{ac}^{t,net} - P_{bat}^{t} + P_{dev,g}^{t} \tag{8.18}$$

Here, P_{bat}^{t} is the cumulative power exchanged with all the BESS and $P_{dev,g}^{t}$ represents the deviation of the load from the forecasted value after implementing DR. The battery power exchange can have both positive and negative values. In (8.18) positive values indicate battery discharging while charging operation is indicated by negative values. The modelling and constraints of the battery energy storage system have been explained in section 8.5.1.2.

8.5.2.2 SPV System Modelling

The calculated value of the SPV output is obtained from the forecasted solar irradiation (8.6). The solar irradiance data is available with a time granularity of one minute. The average value of irradiance taken over 15 minutes is used for each time block [29].

$$P_{cal}^{t,pv} = \begin{cases} P_{r}^{pv}\left(\dfrac{(R^{t})^{2}}{R_{st} \times R_{c}}\right) & \forall\ 0 < R^{t} < R_{c} \\ P_{r}^{pv}\left(\dfrac{R^{t}}{R_{st}}\right) & \forall\ R^{t} > R_{c} \end{cases} \tag{8.19}$$

Here, P_{r}^{pv} is the equivalent generation rating of the SPV system, R^{t} which represents forecasted solar irradiance in W/m² and R_{st} is the irradiance under standard atmospheric conditions and has been taken as 1000 W/m². R_{c} is the radiation point and its value has been assumed as 250 W/² [29]. To account for the effect of atmospheric conditions and inverter efficiency on the generation, the output computed in (8.19) is multiplied with an empirical factor of 0.8.

8.5.3 Results and Discussions

8.5.3.1 Case Study I: Peak Load and PAR Reduction Using Batteries with DR

In this section, the effect of centralized control of active power dispatch and consumption of consumer side batteries on the net load demand has been discussed. Figure 8.4 shows the net load curve with and without the battery power support through DR action. Due to the presence of SPV generation, the net load demand becomes considerably low during daytime and net demand starts increasing from the afternoon onwards. This is because of the combined effect of increased load demand in the evening hours and non availability of solar electricity generation.

Considering the actual load profile, the peak load demand is 1770 kW (time slot 4) and the minimum load demand is 507.9 kW (time slot 37). There is a large difference between demand at the peak point and load at the lowest valley point; and the corresponding PAR value is 1.535. The results reveal that with the application of the energy storage DR, the peak demand is reduced to 1497 kW and the minimum net drawal is raised to 809.7 kW. It is indicated that the power drawal from the grid during peak hours gets reduced, while during low load periods, it is increased.

Figure 8.5 shows the cumulative power exchange with the batteries. Here, the positive values of power exchange indicate battery charging, and negative values indicate energy dispatch from the batteries. The energy is

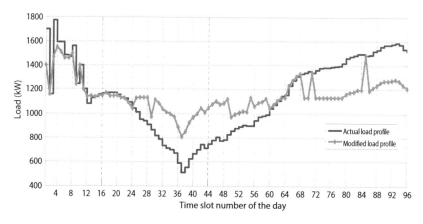

Figure 8.4 Effect of DR on the net load profile.

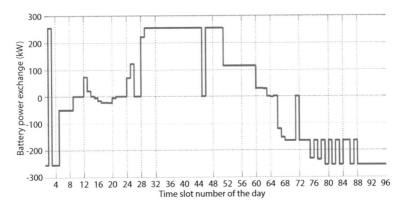

Figure 8.5 Battery charging/discharging cycles over the day.

transferred to the batteries in the time slots where the net demand is lesser and it is released when the demand rises.

During the daytime when sufficient solar energy is available, the energy is transferred to the batteries participating in DR. In the evening hours when the net demand rises, the energy is optimally dispatched from the BESS. This helps in flattening the net load curve by reducing the system peak load, and the peak-to-average ratio also gets improved from 1.535 to 1.302. Thus, the integration of available consumer batteries.

8.5.3.2 Case Study II: Minimizing Load Profile Variations Using Batteries with DR

The forecasted and actual net load profiles are shown in Figure 8.6. It indicates that there have been considerably large deviations of the actual and predicted net load profiles during certain time intervals. It is indicated that during the time slots 32 to 37 and 47 to 53 the difference between the predicted and actual net load demands has been quite large.

Figure 8.7 shows the effect of battery energy storage DR on the net load profile variations from the scheduled demand. The figure shows the load profile variations with and without DR. The cumulative power exchange with the batteries is also shown in this figure. Here the positive values of battery power indicated charging operation, while the negative sign indicates that the energy is being dispatched from the batteries to supply the load. In the time intervals, where the actual demand is more than the forecasted load, the stored energy in the batteries is supplied to the system; and the energy is transferred to the batteries when the demand is less than the forecasted values.

It is clear from the results that the proposed DR scheme has effectively balanced the load profile variations. The variations have been reduced to zero over the larger part of the day; while in the intervals with higher load excursions (31 to 37, 47 to 52, and 65 to 66) these have been considerably reduced. Thus, as revealed in Figure 8.8, with the application of battery energy storage in DR, the actual net load curve follows the forecasted load curve quite closely.

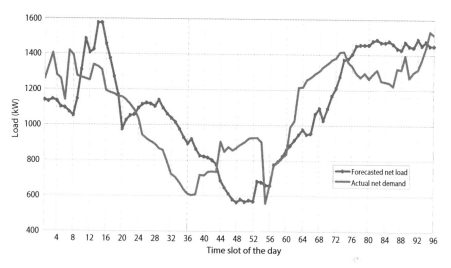

Figure 8.6 Comparison of forecasted and actual net load curves.

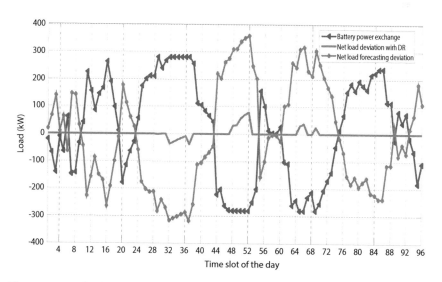

Figure 8.7 Net load variations with/without DR and BESS power exchange.

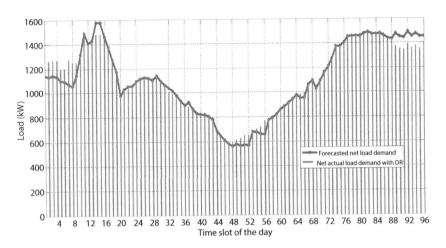

Figure 8.8 Net actual load curve with and without DR.

8.6 Conclusion

Demand response resources can play a vital role in modern power system operations. Particularly, demand response participation of consumer-owned battery energy storage systems offers great potential without causing inconvenience to the consumers. Because of bidirectional power flow and fast response time, the battery energy storage DR provides flexibility to the system, especially in renewable integrated systems. The deployment of residential consumers' battery energy storage for centralized DR helps in reducing the system peak demand and improves peak to average ratio considerably. The battery energy storage DR is also effective in reducing the net variations of the actual load profile from the predicted load profile. Thus, the DR participation of consumers' energy storage systems not only helps in utilizing the SPV power more efficiently but also helps in mitigating the SPV generation variability quite effectively without needing to install additional storage capacity.

References

1. V. S. K. V. Harish and A. Kumar, "Demand side management in India: Action plan, policies and regulations," *Renew. Sustain. Energy Rev.*, vol. 33, pp. 613–624, 2014, doi: 10.1016/j.rser.2014.02.021.

2. A. F. Meyabadi and M. H. Deihimi, "A review of demand-side management : Reconsidering theoretical framework," *Renew. Sustain. Energy Rev.*, vol. 80, May, pp. 367–379, 2017, doi: 10.1016/j.rser.2017.05.207.
3. K. Kostková, P. Kyčina, and P. Jamrich, "An introduction to load management," *Electric Power Systems Research*, vol. 95, pp. 184–191, 2013, doi: 10.1016/j.epsr.2012.09.006.
4. M. T. Johansson and P. Thollander, "A review of barriers to and driving forces for improved energy efficiency in Swedish industry – Recommendations for successful in-house energy management," *Renew. Sustain. Energy Rev.*, vol. 82, April 2017, pp. 618–628, 2018, doi: 10.1016/j.rser.2017.09.052.
5. M. Uddin et al., "A review on peak load shaving strategies," *Renew. Sustain. Energy Rev.*, vol. 82, February 2017, pp. 3323–3332, 2018, doi: 10.1016/j.rser.2017.10.056.
6. I. P. Panapakidis, "A Novel Demand Side Management Strategy Implementation Utilizing Real-Time Pricing Schemes," 2017.
7. T. Boßmann and E. J. Eser, "Model-based assessment of demand-response measures — A comprehensive literature review," *Renew. Sustain. Energy Rev.*, vol. 57, pp. 1637–1656, 2016, doi: 10.1016/j.rser.2015.12.031.
8. G. Strbac, "Demand side management: Benefits and challenges," *Energy Policy*, vol. 36, no. 12, pp. 4419–4426, 2008, doi: 10.1016/j.enpol.2008.09.030.
9. R. Sharifi, S. H. Fathi, and V. Vahidinasab, "A review on Demand-side tools in electricity market," *Renewable and Sustainable Energy Reviews*, vol. 72, December 2016. pp. 565–572, 2017, doi: 10.1016/j.rser.2017.01.020.
10. N. Kinhekar, N. P. Padhy, F. Li, and H. O. Gupta, "Utility Oriented Demand Side Management Using Smart AC and Micro DC Grid Cooperative," *IEEE Trans. Power Syst.*, vol. 31, no. 2, pp. 1151–1160, 2016, doi: 10.1109/TPWRS.2015.2409894.
11. K. Karunanithi, S. Saravanan, B. R. Prabakar, S. Kannan, and C. Thangaraj, "Integration of Demand and Supply Side Management strategies in Generation Expansion Planning," *Renew. Sustain. Energy Rev.*, vol. 73, January, pp. 966–982, 2017, doi: 10.1016/j.rser.2017.01.017.
12. Z. Wang, C. Gu, F. Li, P. Bale, and H. Sun, "Active demand response using shared energy storage for household energy management," *IEEE Trans. Smart Grid*, vol. 4, no. 4, pp. 1888–1897, 2013, doi: 10.1109/TSG.2013.2258046.
13. N. K. Kandasamy, K. J. Tseng, and S. Boon-Hee, "Virtual storage capacity using demand response management to overcome intermittency of solar PV generation," *IET Renew. Power Gener.*, vol. 11, no. 14, pp. 1741–1748, 2017, doi: 10.1049/iet-rpg.2017.0036.
14. J. Vishnupriyan and P. S. Manoharan, "Multi-criteria decision analysis for renewable energy integration: A southern India focus," *Renew. Energy*, vol. 121, pp. 474–488, 2018, doi: 10.1016/j.renene.2018.01.008.
15. L. Ni, F. Wen, W. Liu, J. Meng, G. Lin, and S. Dang, "Congestion management with demand response considering uncertainties of distributed generation

outputs and market prices," *J. Mod. Power Syst. Clean Energy*, vol. 5, no. 1, pp. 66–78, 2017, doi: 10.1007/s40565-016-0257-9.
16. W. Liu, Q. Wu, F. Wen, J. Østergaard, and S. Member, "Day-Ahead Congestion Management in Distribution Systems through Household Demand Response and Distribution Congestion Prices," *IEEE Trans. Smart Grid*, vol. 5, no. 6, pp. 1–10, 2014.
17. C. Zhang, Y. Ding, N. C. Nordentoft, P. Pinson, and J. Østergaard, "Flech: A Danish market solution for DSO congestion management through DER flexibility services," *J. Mod. Power Syst. Clean Energy*, vol. 2, no. 2, pp. 126–133, 2014, doi: 10.1007/s40565-014-0048-0.
18. K. Dehghanpour and S. Afsharnia, "Electrical demand side contribution to frequency control in power systems: a review on technical aspects," *Renew. Sustain. Energy Rev.*, vol. 41, pp. 1267–1276, 2015, doi: 10.1016/j.rser.2014.09.015.
19. M. Yao, J. L. Mathieu, and D. K. Molzahn, "Using Demand Response to Improve Power System Voltage Stability Margins."
20. P. Gupta and Y. P. Verma, "Voltage profile improvement using demand side management in distribution networks under frequency linked pricing regime," *Appl. Energy*, vol. 295, May, p. 117053, 2021, doi: 10.1016/j.apenergy.2021.117053.
21. N. Good, K. A. Ellis, and P. Mancarella, "Review and classification of barriers and enablers of demand response in the smart grid," *Renew. Sustain. Energy Rev.*, vol. 72, November 2016, pp. 57–72, 2017, doi: 10.1016/j.rser.2017.01.043.
22. B. Li, J. Shen, X. Wang, and C. Jiang, "From controllable loads to generalized demand-side resources: A review on developments of demand-side resources," *Renew. Sustain. Energy Rev.*, vol. 53, pp. 936–944, 2016, doi: 10.1016/j.rser.2015.09.064.
23. A. K. Singh and S. K. Parida, "A review on distributed generation allocation and planning in deregulated electricity market," *Renew. Sustain. Energy Rev.*, vol. 82, August 2017, pp. 4132–4141, 2018, doi: 10.1016/j.rser.2017.10.060.
24. S. Dhundhara, Y. P. Verma, and A. Williams, "Techno-economic analysis of the lithium-ion and lead-acid battery in microgrid systems," *Energy Convers. Manag.*, vol. 177, September, pp. 122–142, 2018, doi: 10.1016/j.enconman.2018.09.030.
25. P. Gupta and Y. P. Verma, "Optimisation of deviation settlement charges using residential demand response under frequency-linked pricing environment," *IET Gener. Transm. Distrib.*, vol. 13, no. 12, pp. 2362–2371, 2019, doi: 10.1049/iet-gtd.2018.7116.
26. S. Dhundhara and Y. P. Verma, "Application of micro pump hydro energy storage for reliable operation of microgrid system," vol. 14, pp. 1368–1378, 2020, doi: 10.1049/iet-rpg.2019.0822.
27. M. Yekini Suberu, M. Wazir Mustafa, and N. Bashir, "Energy storage systems for renewable energy power sector integration and mitigation of

intermittency," *Renew. Sustain. Energy Rev.*, vol. 35, pp. 499–514, 2014, doi: 10.1016/j.rser.2014.04.009.
28. A. M. Vega, F. Santamaria, and E. Rivas, "Modeling for home electric energy management: A review," vol. 52, pp. 948–959, 2015, doi: 10.1016/j.rser.2015.07.023.
29. P. Gupta and Y. P. Verma, "Role of Residential Demand Response in Optimizing Utility Cost in PV Integrated System under Frequency Linked Pricing Environment," 2019.

9

Impact of Battery Energy Storage Systems and Demand Response Program on Locational Marginal Prices in Distribution System

Saikrishna Varikunta* and Ashwani Kumar

Department of Electrical Engineering, National Institute of Technology, Kurukshetra, Haryana, India

Abstract

The pricing of electricity consumption is an important aspect of power system economics. Electricity pricing based on the Locational Marginal Price (LMP) is gaining attention in electricity markets all around the world due to the better distribution system markets and energy-efficient distribution of electricity with hybrid combination of energy resources. The pricing should be such that the generation, transmission, and end users should all benefit. The fairness of pricing should be present in the pricing scheme. By using Locational Marginal Price (LMP), active and reactive power pricings can be made. With LMP-based pricing, the generator, transmission operator and end users are fairly benefited. It is also believed that with the knowledge of the LMPs end users can manage their load and try to participate in a Demand Response (DR) program.

In this chapter, the Locational Marginal Price (LMP) of active and reactive powers is obtained considering the role of storage devices and the impact of a demand response program on their variations. The impact of installation of Shunt capacitor, Energy Storage System (ESS) like Battery Energy Storage System (BESS) and Load participation in Demand Response (DR) program on the LMPs is discussed. Based on the LMPs, the technical and economic aspects are studied. The OPF problem has been solved using General Algebraic Modeling System (GAMS) software and the results are plotted using MATLAB.

---

*Corresponding author: varikuntasaikrishna17@gmail.com

Sandeep Dhundhara, Yajvender Pal Verma, and Ashwani Kumar (eds.) *Energy Storage Technologies in Grid Modernization*, (239–282) © 2023 Scrivener Publishing LLC

Keywords: Battery energy storage systems, demand response program, locational marginal prices, microgrids

9.1 Introduction

Today's power system network is very complex and there is a paradigm shift in its operation from conventional to automatic due to the amalgamation of renewable energy sources with conventional power plants. The operation of the power sector is now competitive with wholesale selling and buying of the power in a pool and bilateral mode. The power is being sold at a competitive price in the electricity markets. The transparent pricing mechanisms are gaining importance for fair and transparent operation of the power sector where locational marginal prices are essential to computing prices of the electricity. Locational Marginal Pricing (LMPs) represents the cost of buying and selling power at different locations within wholesale electricity markets, usually called Independent System Operators (ISOs). LMPs give information about the three components, Energy Price, Congestion Cost, and Losses [1]. The LMPs give fair information of cost components and information about the network operation and behavior of load. Many authors have done research in this area of obtaining the cost. Both real and reactive power pricing is essential to establish what consumers pay for their usage of electricity.

With the emergence of smart grid, many new components are being added to the network like battery energy storage systems and renewable energy sources, and demand-side management is also playing a key role in the operation of the system. A brief overview of the battery energy storage systems and demand-side management follows.

9.1.1 Battery Energy Storage System (BESS)

Energy storage is a bridge between intermittent sources and fluctuating loads. The energy produced at one moment can be used later by storing it. One type of energy storage is electricity storage. Mechanical, electrochemical (or batteries), thermal, electrical, and hydrogen storage methods are some of the major technologies that fall under the umbrella of energy storage technologies.

A Battery Energy Storage System (BESS) is a type of energy storage system (ESS) that collects energy from various sources, and stores it in batteries (rechargeable type) for subsequent utilization. Different BESS-Battery types are Li-Ion batteries, Lead-Acid (PbA) batteries, Nickel-Cadmium (Ni-Cd) batteries, Sodium sulfur (Na-S) batteries and Flow batteries, etc.

Different components in the Battery Energy Storage System (BESS) are the Energy management system (EMS), the Battery system, and Power Conversion System (PCS). The battery system includes the battery pack, battery management system (BMS), and battery thermal management system (B-TMS). The cells are protected by BMS and the cell temperature is controlled by B-TMS. EMS will ensure the safety and reliability of the system overall operation. PCS will monitor and control various power electronic components which are used in the power conversion (Bi-directional) between the grid and the battery.

Energy storage device applications vary depending on the time needed to connect to the generator, transmitter, and place of use of energy, and on energy use.

Some of the BESS applications in Grid-Utility are given below with different size ranges [12]:

- *Electric energy time-shift [1-500 MW]:*
  Electric energy time-shifting is acquiring affordable electric energy during periods when prices or system marginal costs are low in order to charge the storage system and consume or sell the stored energy at a later time when prices or costs are high.
- *Electric Supply Capacity [1-500 MW]:*
  Energy storage may be used to postpone or lessen the requirement to purchase capacity in the wholesale electricity market, based on the situation in a particular electric supply system.
- *Regulation [10-40 MW]:*
  Among the auxiliary services, regulation is one of those for which storage is particularly well suited.
- *Spinning, non-spinning, and supplemental reserves [10-100MW]:*
  When a portion of the usual sources of electric supply is suddenly unavailable, an electric grid must have reserve capacity that may be used. BESS will act as reserve capacity,
- *Voltage Support [1-10MVAr]:*
  Reactive power control can be carried out in addition to active power management by a battery energy storage system (BESS) outfitted with an appropriately sophisticated inverter (Power Conversion Systems). This enables the deployment of a battery energy storage system in a grid to additionally assist the grid with reactive power and adjust the power factor of loads.

- **Black Start [5-50 MW]:**
  Black start refers to the power system's capacity to rebound from a blackout by restarting selected components. In order to create an interconnected system once more, disconnected power plants must be started separately and progressively rejoined. When a blackout occurs and the grid needs to be reset from scratch, then BESS is will provide start-up to the power plants such that they can be brought back on-line.
- **Load following [1-100 MW]:**
  A BESS that is the appropriate fit may also offer longer-duration services, such as load-following and ramping services, to make sure supply keeps up with demand. To make sure supply and demand are balanced, a properly scaled BESS may also offer longer-duration services like load-following and ramping services.
- **Demand Charge Management [50 KW-10 MW]:**
  End users can employ energy storage systems to lower their total expenses for energy consumption by lowering their demand during peak hours which are set by the utility. BESS will be able to compensate for the reduced demand during peak hours; thereby the load burden on the grid utility decreases and so not only the end users but also the utility grid will benefit.
- Others applications and benefits are: Transmission and Distribution (T&D) upgrade deferral, Transmission congestion management, power quality, etc.

9.1.2 Demand Response Program

The demand of different consumers may change its patterns as per the operating conditions in a system that may respond to changes in the price of electricity during a whole day. Such demand may have the incentive to reduce their demand during the peak loads and may help the reduction and valley shaving during the peak hours of load demand. This will help to have payments that may be designed to lower electricity use at times of high wholesale market prices. The demand response has the capability of changing the electric usage by the customers/end users from their normal consumption patterns as per the response to changes in the price of electricity during the whole day or in real time and can have the mechanisms for incentive payments that are designed to lower electricity usage during the time of high wholesale market prices or when system security and reliability is threatened [13]. The Demand response programs are classified into two categories: Dispatchable (or) Incentive-based programs and

Non-Dispatchable (or) Time-Based programs. Some of the Incentive-Based demand response programs are Direct control, Interruptible/Curtailable, Demand binding, Emergency market, capacity market, ancillary services market type programs, etc. Some of the Time-Based demand response programs are Time-of-Use, Critical Peak Pricing, Extreme Day CPP, Extreme Day pricing, and Real-Time pricing programs, etc.

Some of the Benefits of DR program are:

- The bill savings and incentive payments received by consumers who alter their power usage in response to time-varying electricity tariffs or incentive-based programs are referred to as participant financial benefits.
- The load shape can be altered.
- The customer has control over their consumption, and Easy to implement.
- With the DR program the reduction of outages in the system is observed, thereby reliability benefits such as operational safety and sufficiency savings can be achieved.
- At distribution level the benefits of DR program implementation are: shifting and reducing the peak demand, meeting future energy needs, postponement of distribution system capital investment, voltage-related issues are resolved, reduction of congestion at the substation and making integration of renewable energy resources simple, etc.
- At the transmission level, the benefits of DR program are reduction of transmission congestion, postponement of transmission expansion projects and improvement of transmission network reliability, etc.
- At the generation level the benefits due to DR are lower generating equipment's operational costs, etc.

Therefore, it is essential to find the impact of storage devices and the role of demand-side management on the LMPs. Reference [2], discussed the important role of LMPs in restructured wholesale power markets. The work reported by the authors extensively presented and analysed various AC and DC optimal power flow models to better understand the calculation of LMPs. In [3], the authors proposed active and reactive power pricing on the basis of marginal costs. In [4] the real-time reactive power pricing is discussed by considering the price depended on demand function; the author took minimization of active power generation cost as the objective function. In [5], the authors included the reactive power production cost along with the active power production cost in objective function; the Optimal Power Flow (OPF) problem is solved using the sequential quadratic programming (SQP) method, and various factors affecting the reactive power LMPs are discussed. In [6] the optimal reactive power management is discussed and the OPF objective function is to minimize the cost of reactive power and it is solved by using Particle Swarm Optimization (PSO) algorithm. In [7] a new approach for reactive power pricing and the cost allocation of reactive power is proposed and a quadratic cost function for reactive power support is calibrated. In [8] the authors used a modified OPF with the objective function of minimizing both the real and reactive power from which LMPs of real and reactive power has been obtained; they also discussed the significance of negative LMPs and showed some finding regarding the nature of LMP during Off-peak load condition.

The LMPs are obtained by solving the OPF problem by using GAMS programming in [9]. Five different cases are considered in the analysis and for each case the changes in LMPs, their technical and economic aspects are discussed. Based on the LMPs, the Total Revenue generated, Congestion Cost, Revenue of Generator and Profit of Generator are calculated for each case. How the LMPs impacts on economic aspects of generator is also discussed.

The five case studies are: **Case 1**, the objective function is the minimization of the active power production cost. In **Case 2**, the reactive power production cost is included in the objective function of case 1. In **Case 3**, a shunt capacitor is considered and its annual objective function is the modification of case 2 by including the capacitor cost. In **Case 4**, battery energy storage devices are considered along with shunt capacitor, and the objective function consists of the objective function of case 3 including the BESS cost. In Case 5, the LMPs' applications are studied, like the load is assumed to participate in the Demand Response (DR) program and incentives are provided accordingly for curtailment of load based on the LMPs [10, 11]; the objective function consists of the objective function of case 4 including the utility maximization term. Later in this chapter, due to the installation shunt capacitor, how the reactive power LMPs will change is discussed and due to energy storage devices and load curtailment how the LMPs of active power will change are discussed. An optimization problem has been solved using GAMS programming with CONOPT-NLP solver [15–17].

9.2 Problem Formulation and Solution Using GAMS

The Locational Marginal Price (LMP) is defined as the marginal cost of supplying the next increment of electric energy at a certain location while

Impact of BESS and DRP on LMP in Distribution System

taking into account the generation marginal cost and the physical features of the transmission system [14]. The LMPs of Active and Reactive power are obtained by solving the OPF problem, in which the objective function is minimized subject to equality and inequality constraints. In this chapter, the OPF problem in solved using GAMS programming with CONOPT-NLP solver [15–17]. In this section the various objective functions for different cases and constraints are mentioned.

The OPF in this chapter is formulated for five case studies. In case 1, the objective function is minimization of the active power production cost and it is given by equation (9.1). In case 2, the objective function is to minimize both active and reactive power production cost, without a shunt capacitor in the system and it is given by the equation (9.4). In case 3, the objective function of case 2 is modified by adding the Capacitor investment cost to it; here a variable shunt capacitor is placed at bus 4 in the system. In case 4, the modification in the system is an Energy Storage Device which is added along with the Shunt Capacitor at bus 4; here objective function for this case is obtained by modifying objective function of Case 3 by adding the BESS investment cost (per day) and the operation cost, In case 5, the load at bus 2 is assumed to be participated in DR program and the load that bus is curtailed and incentives are paid accordingly, and the objective function is obtained by modifying case 4 by adding utility maximization function, for this case is the modification of case 4.

9.2.1 Objective Functions for Case Studies: Case 1 to Case 5

9.2.1.1 Case 1: Is Minimization of the Active Power Production Cost

The Objective function of the OPF is represented as

$$Min\ C_1 = \sum_{i \in G} \left(C_{pi}(P_{gi}) \right) \tag{9.1}$$

Where C_1 is the sum of cost of real power production cost of all generator. $C_{pi}(P_{gi})$ is the cost of production of active power by the ith generator. G is total no. of generators in the system.

The active power production cost can be represented as

$$C_{pi}(P_{gi}) = a_{pi}P_{gi}^2 + b_{pi}P_{gi} + c_{pi} \quad i \in G \tag{9.2}$$

9.2.1.2 Case 2: Minimization of the Active Power Production and Reactive Power Production Cost

The active power production cost is represented in $/hr. The reactive power production cost is often neglected. The reactive production cost is also known as Opportunity cost. During heavy loading conditions the generator will supply the reactive power to the system by reducing some of its active power generation [5]. So, thereby the cost of active power selling due to reactive power production is reduced. The opportunity cost of the generator depends on the real-time dispatch and the price of active power at that instant. Hence, the exact calculation of cost of reactive power production is complex. For simplicity, the cost of reactive power production of ith generator, $C_{qi}(Q_{gi})$ is represented as [18]:

$$C_{qi}(Q_{gi}) = \left[C_{pi}(S_{gi,max}) - C_{pi}\left(\sqrt{(S_{gi,max})^2 - (Q_{gi})^2}\right) \right] \times k \quad (\$/hr) \tag{9.3}$$

Where $S_{gi,max}$ is the nominal apparent power of the generator and k is the profit rate of active power produced, usually its range is [5% – 10%]. Here, it is assumed that $S_{gi,max} = P_{gi,max}$.

The Objective function for Case 2 is represented as:

$$\text{Min } C_2 = \sum_{i \in G} \left(C_{pi}(P_{gi}) + (C_{qi}(Q_{gi})) \right) \tag{9.4}$$

9.2.1.3 Case 3: Minimization of the Active Power Production and Reactive Power Production Cost Along with Capacitor Placement

In this case, a shunt capacitor is placed in the system to provide reactive power support and to improve the voltage profile. The cost function/model of the capacitor investment cost of the jth capacitor is given by the equation (9.5):

$$C_j(Q_{cj}) = \left(\frac{Investment\,Cost\,(\$/MVAr)}{Operating\,period\,(hr) \times u} \right) \times Q_{cj}\,(\$/hr) \quad j \in C \tag{9.5}$$

Impact of BESS and DRP on LMP in Distribution System

Where C is the total number of shunt capacitors present in the system.

Here, the investment cost is considered as $11600/MVAr, the operating period (or) the life span of capacitor is assumed to be 15 years (15 yrs = 15*8760 hrs = 131400 hrs), the usage rate (u) is taken as 2/3, the above Equation can be simplified as:

$$C_j(Q_{cj}) = Q_{cj} \times 13.24 \ (\$/MVAhr) \quad (9.6)$$

Here, Q_{cj} in MVAr's.

The Objective function of this case is obtained by modifying case 2 objective function, and it is given by:

$$\text{Min } C_3 = \sum_{i \in G}\left(C_{pi}(P_{gi}) + \left(C_{qi}(Q_{gi})\right)\right) + \sum_{j \in C}\left(C_j(Q_{cj})\right) \quad (9.7)$$

9.2.1.4 Case 4: Minimization of the Active Power Production and Reactive Power Production Cost Including Capacitor and BESS Cost

In this case, Energy Storage System (ESS) like Battery Energy Storage System (BESS) is considered in a distribution system along with the shunt capacitor.

The State of Charge (SOC) equation of simple ESS/BESS model is given by:

$$SOC_i(t) = SOC_i(t-1) + (P_i^{char}(t)\eta_{char} - P_i^{dis}(t)/\eta_{dis})\,\Delta t \quad (9.8)$$

Where $P_i^{char}(t)$ and $P_i^{dis}(t)$ are Charging and Discharging power of ESS/BESS at time t, η_{char} and η_{dis} are charging and discharging efficiency rates, i is the bus no. and Δt is the time interval length.

The BESS Size is optimally obtained solving the problem using GAMS taking the cost equations in the optimization problem and the cost equations are given (9.9, 9.10) and are refereed in [19, 20]:

The Annual investment cost of BESS:

$$C_{BESS}^{inv} = \left(Z_{BESS}^{ERATED} E_{BESS}^{ERATED} + Z_{BESS}^{PRATED} P_{BESS}^{PRATED}\right) \cdot A_{BESS}$$

$$(9.9)$$

Where

$$A_{BESS} = \frac{r_{in}(1+r_{in})^y}{(1+r_{in})^y - 1} *$$

Z_{BESS}^{ERATED} and Z_{BESS}^{PRATED} are investment cost of the BESS Capacity and Power respectively. E_{BESS}^{ERATED} and P_{BESS}^{PRATED} are the rated BESS Capacity and Power respectively.

A_{BESS} is the coefficient to calculate BESS annual cost. r_{in} and y are interest rate and life time of BESS respectively.

The Operational cost of BESS:

$$C_{BESS}^{Oper} = Z_{BESS}^{Opr} \cdot \sum_{t=1}^{24} \left(P_i^{char}(t) + P_i^{dis}(t) \right) \quad (9.10)$$

where Z_{BESS}^{Opr} is the operational cost of the BESS.

The Objective function of case 4 is given by:

$$\text{Min } C_4 = C_3 + (C_{BESS}^{inv}/365) + C_{BESS}^{Oper} \quad (9.11)$$

9.2.1.5 Case 5: Minimization of the Active Power Production and Reactive Power Production Cost Including Capacitor and BESS Cost and Taking the Impact of Demand Response Program

In this case, the load at any particular bus is assumed to participate in the Demand Response (DR) Program. For curtailed active power demand, the incentives paid accordingly.

The customer cost function [10] at bus i for load curtailment of x_i is

$$C_i = K_1 x_i^2 + K_2 x_i (1 - \theta_i) \quad i \in N \quad (9.12)$$

Here i is assumed to be the bus number of the system and N is the number of buses in the system. Also, here it is assumed to be the load is controllable and can be curtailed by the Utility Operator. θ_i is the willingness of the customer to participate in the DR program and its range is [0 – 1].

K_1 and K_2 are the customer cost coefficients that are calibrated by the Utility. Here, for simplicity the constants chosen are, $K_1 = 1/2$, $K_2 = 1$ and $\theta = 0.5$.

Let y_i be the incentive paid for the curtailment of load x_i [11]. The Customer Benefit can be represented as:

$$Customer\ Benefit_i = y_i - C_i \qquad (9.13)$$

The appropriate incentive y_i paid for satisfying the customer and utility will be within the range as given below

$$[K_1 x_i^2 + K_2 x_i (1 - \theta_i)] \leq y_i \leq \lambda_{pi} x_i \qquad (9.14)$$

The term $(y_i - \lambda_{pi} x_i)$ in the Equation (9.15) represents the utility maximization function; here in this DR program both the customer and utility are benefited, i.e., the customer will get more or equal amount for this load curtailment based on their customer benefit function, the utility will pay to the customer less than the maximum interruptible value, i.e., it depends on the value of $\lambda_{pi} x_i$. Here for this case, the LMPs are taken from the previous case, i.e., without DR program case 4. The budget limit on utility can also be placed to restrict the amount of incentives paid.

The objective function for case 5 is given by:

$$Min\ C_5 = C_3 + (C_{BESS}^{inv}/365) + C_{BESS}^{Oper} + \left[y_i - \lambda_{pi} x_i\right] \qquad (9.15)$$

9.2.2 Real and Reactive Power Equality Constraints

The following constraints are considered for the above-mentioned cases:

9.2.2.1 Equality Constraints

The Load Flow Equations are the equality constraints for the OPF problem. The Active and Reactive power balance equations are (at bus i):

$$P_{gi} - P_{di} - \sum_{j \in N} |V_i||V_j||Y_{ij}| \cos(\theta_{ij} + \delta_j - \delta_i) = 0 \qquad (9.16)$$

$$Q_{gi} - Q_{di} + \sum_{j \in N} |V_i||V_j||Y_{ij}|\sin(\theta_{ij} + \delta_j - \delta_i) = 0 \quad (9.17)$$

The change in equality constraints for different cases are:

For Case 3 to Case 5: Reactive power balance equation

$$Q_{gi} + Q_{cj} - Q_{di} + \sum_{j \in N} |V_i||V_j||Y_{ij}|\sin(\theta_{ij} + \delta_j - \delta_i) = 0 \quad (9.18)$$

For Case 4: Real power balance equation

$$P_{gi} + P_i^{dis} - P_{di} - P_i^{char} - \sum_{j \in N} |V_i||V_j||Y_{ij}|\cos(\theta_{ij} + \delta_j - \delta_i) = 0 \quad (9.19)$$

For Case 5: Real power balance equation

$$P_{gi} + P_i^{dis} - P_{di} - P_i^{char} - x_i - \sum_{j \in N} |V_i||V_j||Y_{ij}|\cos(\theta_{ij} + \delta_j - \delta_i) = 0$$

$$(9.20)$$

N is the total number of buses in the system.

9.2.2.2 Inequality Constraints: (at any bus i): Voltage, Power Generation, Line Flow, SOC, Battery Energy Storage Power

Voltage Magnitude Constraints

$$|V_{i,\min}| \leq |V_i| \leq |V_{i,\max}| \quad (9.21)$$

Voltage Angle Constraints

$$\delta_{i,\min} \leq \delta_i \leq \delta_{i,\max} \quad (9.22)$$

Active Power Generation Constraints

$$P_{gi,\min} \leq P_{gi} \leq P_{gi,\max} \quad i \in G \quad (9.23)$$

Reactive Power Generation Constraints

$$Q_{gi,\min} \leq Q_{gi} \leq Q_{gi,\max} \quad i \in G \tag{9.24}$$

Complex (or) Apparent power constraint on generator

$$\sqrt{P_{gi}^2 + Q_{gi}^2} \leq S_{gi,\max} \quad i \in G \tag{9.25}$$

Transmission line constraints are limits

$$|P_{ij}| \leq P_{ij,\max} \quad i \neq j \text{ and } (i,j) \in N \tag{9.26}$$

Reactive power output constraint of shunt capacitor (This Constraint is included for case 3 and case 4 only)

$$0 \leq Q_{cj} \leq Q_{cj,\max} \quad j \in C \tag{9.27}$$

ESS/BESS constraints:
Charging and Discharging constraints:

$$P_{\min}^{char} \leq P_i^{char}(t) \leq P_{\max}^{char} \tag{9.28}$$

$$P_{\min}^{dis} \leq P_i^{dis}(t) \leq P_{\max}^{dis} \tag{9.29}$$

SOC limits:

$$SOC_{\min} \leq SOC(t) \leq SOC_{\max} \tag{9.30}$$

9.2.3 Modified Lagrangian Function

The modified Lagrangian function for the OPF problem (in simplified form) is given by

L = {Objective Function No. k}

$$-\lambda_{pi} * \left[P_{gi} - P_{di} - \sum_{j \in N} |V_i||V_j||Y_{ij}|\cos(\theta_{ij} + \delta_j - \delta_i) \right]$$

$$-\lambda_{qi} * \left[Q_{gi} - Q_{di} + \sum_{j \in N} |V_i||V_j||Y_{ij}|\sin(\theta_{ij} + \delta_j - \delta_i) \right] \quad (9.31)$$

Where k is the Case number. In the above equation (9.31), the Lagrangian function is different for different case. λ_{pi} and λ_{qi} are the Active and Reactive Power Locational Marginal Prices (LMPs) respectively. The Lagrangian multipliers λ_{pi} and λ_{qi} are only considered in equation (9.31), the Lagrangian multiplier's for generation limits, line constraints and voltage limits are not mentioned here even though the conditions are to meet.

9.2.4 Generator Economics Calculations

In this section, we formulate the Total Revenue generated, Congestion Cost, Revenue of Generator and Profit of Generator based on the LMPs [8]. The simple Electricity Market model for the IEEE 5-bus is shown in Figure 9.1, and the power and Money/cash flows can be seen. In Figure 9.1, it can be seen that the power flow starts from generation center to load center; in special cases when renewable energy sources and storage devices included at load center then the power can exchange between load and Transmission system operator (TSO). Here in this chapter as the BESS is added, during discharging period the power from BESS will satisfy the load and it gets paid from TSO. When Load participates in DR program the Load will get paid incentives according to their curtailment by TSO. The revenue is generated from the load is collected by the TSO and TSO will pay to the GENCOs by charging some amount for the congestion if any. The revenue obtained by the GENCOs also includes the cost of the loss.

The total revenue generated by the loads is

$$\text{Total Revenue generated} = \sum_{i \in N} [\lambda_{pi} P_{di} + \lambda_{qi} Q_{di}] \quad (9.32)$$

Impact of BESS and DRP on LMP in Distribution System 253

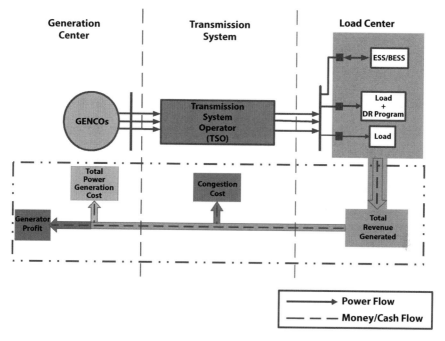

Figure 9.1 Electricity Market Structure for 5-bus system.

The Congestion cost is

$$C_{P,Congestion} = \frac{1}{2} \sum_{\substack{i,j \in N \\ i \neq j}} \left| \left[\lambda_{pi} - \lambda_{pi} \right] * P_{i,j} \right| \quad (9.33)$$

$$C_{Q,Congestion} = \frac{1}{2} \sum_{\substack{i,j \in N \\ i \neq j}} \left| \left[\lambda_{pi} - \lambda_{pi} \right] * Q_{i,j} \right| \quad (9.34)$$

$$\text{Total Congestion Cost} = \sqrt{C_{P,Congestion}^2 + C_{Q,Congestion}^2} \quad (9.35)$$

In, Equation (9.35) the $C_{Q,Congestion}$ is very small value comparted to the $C_{P,Congestion}$, so an approximation for the Total Congestion Cost is given by

$$\text{Total Congestion Cost} \approx C_{P,Congestion} \quad (9.36)$$

Revenue of Generator is:

$$\text{Revenue of Generator} = \text{Total Revenue generated} - \text{Total Congestion cost} \quad (9.37)$$

Note that here the revenue of generator includes the cost of losses. Profit of Generator is:

$$\text{Profit of Generator} = \text{Generator Revenue} - \text{Cost of Generation} \quad (9.38)$$

9.3 Case Study: Numerical Computation

A Standard IEEE 5-bus system [21] is considered for the case study and its single line diagram is given in Figure 9.2.

In the IEEE 5-Bus system diagram, for case 3 the shunt capacitor switch is on, for case 4 both the shunt capacitor and BESS switch are closed and

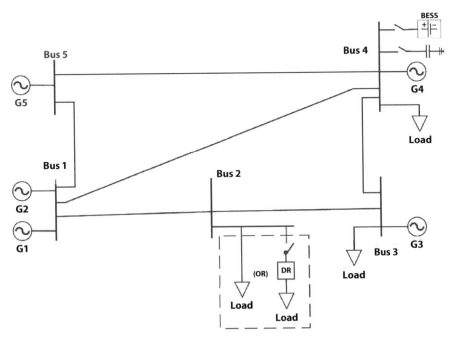

Figure 9.2 Single Line diagram of IEEE 5 bus system.

Table 9.1 Generator power constraints and cost co-efficient data.

Bus no.	Generator no.	P_{min} (MW)	P_{max} (MW)	Q_{min} (MVAr)	Q_{max} (MVAr)	b ($/MWh)
1	1	0	40	-30	30	14
1	2	0	170	-127.5	127.5	15
3	3	0	520	-390	390	30
4	4	0	200	-150	150	40
5	5	0	600	-450	450	10

Table 9.2 Line data.

From bus	To bus	R (pu)	X (pu)	B/2 (pu)	Limit (MVA)
1	2	0.00281	0.0281	0.00356	400
1	4	0.00304	0.0304	0.00329	400
1	5	0.00064	0.0064	0.01563	400
2	3	0.00108	0.0108	0.00926	400
3	4	0.00297	0.0297	0.00337	400
4	5	0.00297	0.0297	0.00337	240

for case 5 the load at bus 2 with DR program is used instead of only load. *Generator Power Constraints and Cost co-efficient Data* are given in Table 9.1. The Line data for the test case is given in Table 9.2.

The Typical load profile for the test case is considered as shown in Figure 9.3. The load $(P_{di} + jQ_{di})$ at each *bus i* is multiplied with the load factor, which is obtained from the Load profile for time period of 24 hrs. The load during peak hr, i.e., 18th hr at each bus is given in Table 9.3. The ESS/BESS data is given in Table 9.4.

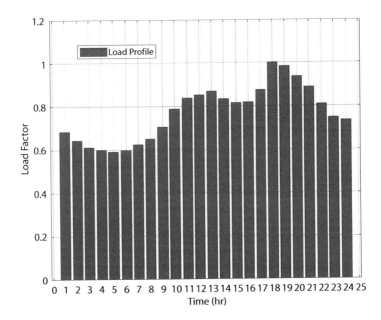

Figure 9.3 Load profile.

Table 9.3 Load at peak hour at each bus.

Bus no.	Pd (MW)	Qd (Mvar)
1	0	0
2	300	98.61
3	300	98.61
4	400	131.47
5	0	0

Table 9.4 ESS/BESS data [17].

Parameter	Value
E_{BESS}^{ERATED}	400 MWh
P_{BESS}^{ERATED}	50 MW
η_{char}	0.89
η_{dis}	0.91
r_{in}	0.08
Y	8 yrs.
Z_{BESS}^{Opr}	0.005 ($/hr)
E_{BESS}^{ERATED}	100 ($/kWh)
P_{BESS}^{ERATED}	250 ($/kW)

9.4 Results and Discussions

9.4.1 Case 1: Minimization of the Active Power Production Cost

For Case 1, the Demand and Generation of Active and Reactive powers are plotted in Figure 9.4a. The hourly active and reactive power loss and the

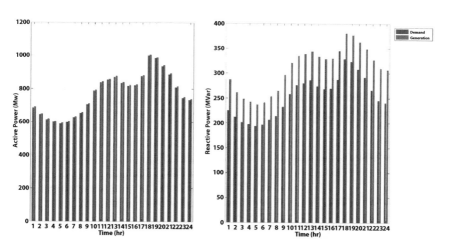

Figure 9.4a Case 1: Active and reactive power demand and generation.

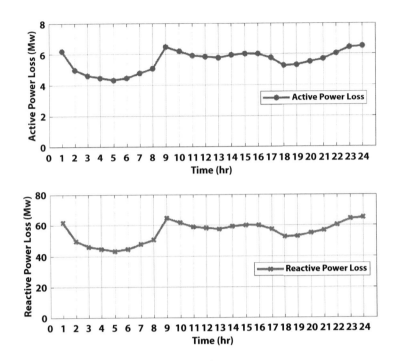

Figure 9.4b Case 1: Active and reactive power loss.

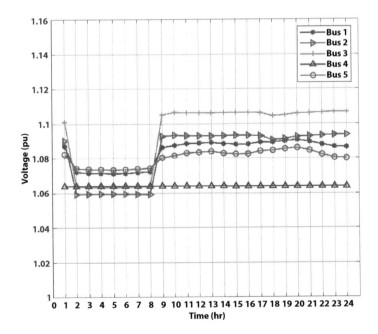

Figure 9.4c Case 1: Voltage profile.

voltage profile at different bus are plotted in Figure 9.4b and Figure 9.4c, respectively.

The Active and Reactive Power LMPs at different bus for 24 hrs. period are plotted in Figure 9.4d. Here, the observation is that, LMPs of reactive power at some buses are negative.

The significance of *"Negative λ_q"*: It signifies that the load/generator which absorb reactive power at that bus will be paid compensation money and the generator which inject the reactive power at that bus will be penalized.

The Generator Economics calculations are calculated and the results are plotted in Figure 9.4e. In Figure 9.4e it can be seen that at 5th hr. the generator profit is small and negative; also, in the Zoom plot it is shown that the

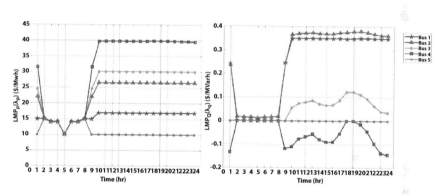

Figure 9.4d Case 1: LMPs of active and reactive power at different buses.

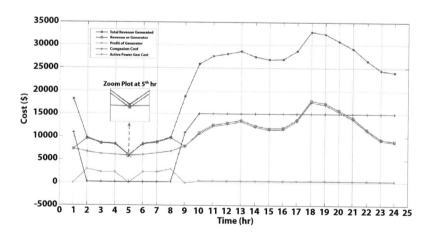

Figure 9.4e Case1: Generator economics: Cost and profit.

generator revenue is less than the cost of active generation and it is due to the line congestion cost.

9.4.2 Case 2: Minimization of the Active Power Production and Reactive Power Production Cost

The Active and Reactive power Demand and Generation graph is shown in Figure 9.5a.

For case 2, the hourly active and reactive power loss and the voltage profile at different bus are plotted in Figure 9.5b and Figure 9.5c, respectively. Compared to the voltage profile of case 1 in Figure 9.4c, the voltage profile obtained in this case is improved.

For Case 2, the Active and Reactive Power LMPs at different bus for 24 hrs period are plotted in Figure 9.5d. Here, the observation is that, LMPs of reactive power at all buses are positive while compared to the LMP's obtained by case 1 (see in Figure 9.4d). So, in this case there are no negative reactive power LMPs.

For Case 2, Generator Economics calculations are calculated and the results are plotted in Figure 9.5e. In Figure 9.5e, at 5th hr. the generator profit is negative but compared to that in the Figure 9.4e, the value is less negative, i.e., generator incurring loss is less, and also at 5th hr. the cost of generation is greater than the generator revenue.

The important findings by the case 2 results are the LMPs of reactive power at all buses are positive and the voltage profile is slightly improved.

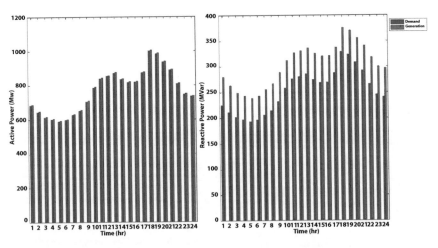

Figure 9.5a Case 2: Active and reactive power demand and generation.

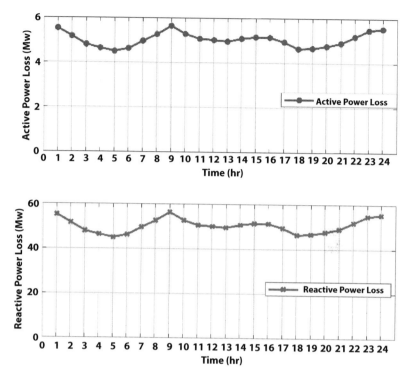

Figure 9.5b Case 2: Active and reactive power loss.

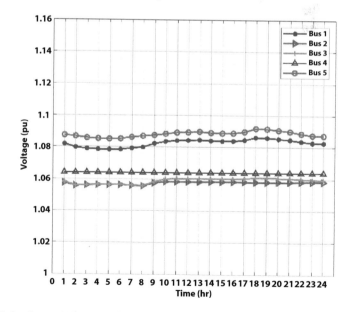

Figure 9.5c Case 2: Voltage profile.

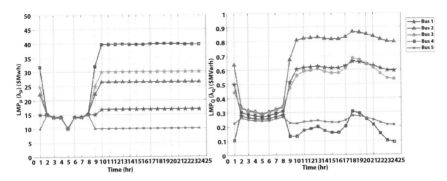

Figure 9.5d Case 2: LMPs of active and reactive power at different buses.

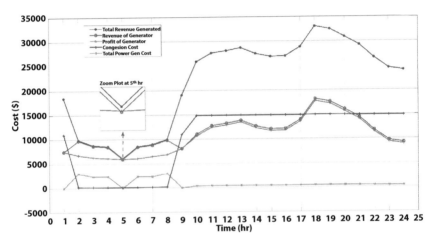

Figure 9.5e Case 2: Generator economics: cost and profit.

9.4.3 Case 3: Minimization of the Active Power Production and Reactive Power Production Cost Along

The Active and Reactive power Demand and Generation and the Capacitor size is shown in Figure 9.6a.

Due to the shunt capacitor placement at bus 4, the voltage profile at bus 4 is further improved. Here, the capacitor is considered to be continuously variable over 24 hr.

For Case 3, the hourly active and reactive power loss and the voltage profile at different bus are plotted in Figure 9.6b and Figure 9.6c, respectively. Compared to the voltage profile of case 1 in Figure 9.4c, the voltage profile obtained in this case is improved.

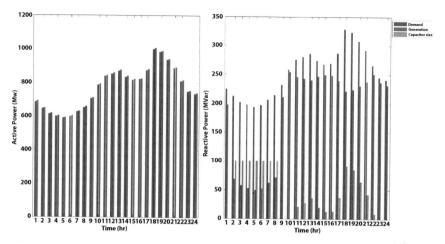

Figure 9.6a Case 3: Active and reactive power demand and generation and capacitor size.

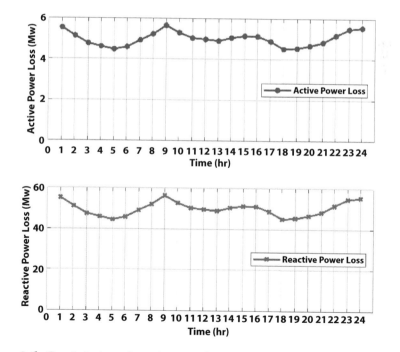

Figure 9.6b Case 3: Active and reactive power loss.

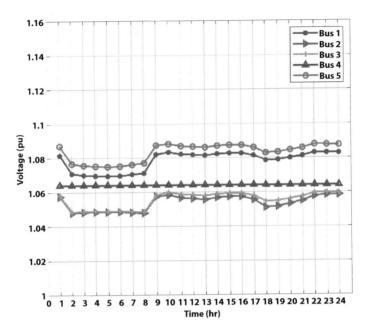

Figure 9.6c Case 3: Voltage profile.

The Capacitor size and its cost at each hour is shown in Figure 9.6d.

For Case 3, the Active and Reactive Power LMPs at different bus for 24 hrs period are plotted in Figure 9.6e. Here, the observation is that, LMPs of reactive power at all buses are positive while compared to the LMPs

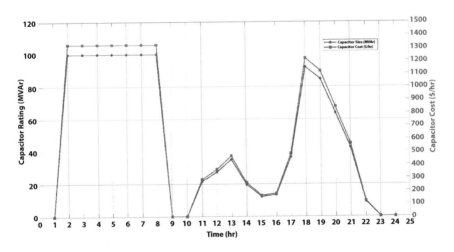

Figure 9.6d Case 3: Capacitor size and cost vs. time.

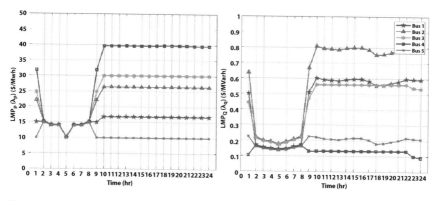

Figure 9.6e Case 3: LMPs of active and reactive power at different buses.

obtained by case 1 (see in Figure 9.4d). So, in this case there are no negative reactive power LMPs.

For Case 3, Generator Economics calculations are calculated and the results are plotted in Figure 9.6f. In Figure 9.6f, at 5th hr the generator profit is negative but compared to that in the Figure 9.4e, the value is less negative, i.e., generator incurring loss is less, and also at 5th hr the cost of power generation is greater than the generator revenue. The important findings by the case 3 results are the LMPs of reactive power at all buses are positive, the reactive power production cost is reduced, active and reactive power losses are reduced and voltage profile is improved.

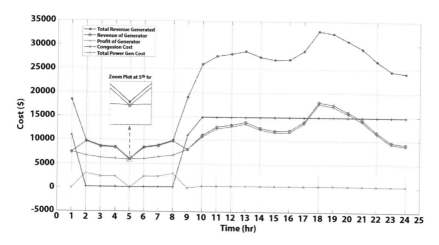

Figure 9.6f Case 3: Generator economics: cost and profit.

9.4.4 Case 4: Minimization of the Active Power Production and Reactive Power Production Cost

In this case, a ESS/BESS is added along with the shunt capacitor at bus 4, and the objective function of this case is mentioned in Section 9.2. The ratings of ESS/BESS are given in Table 9.4. The Active and Reactive power Demand and Generation and the Capacitor size is shown in Figure 9.7a.

Due to the ESS/BESS installation the change in active power LMPs is observed. The shunt capacitor will make change in the reactive power LMPs; also the voltage profile at bus 4 is improved.

For case 4, the hourly active and reactive power loss and the voltage profile at different bus are plotted in Figure 9.7b and Figure 9.7c respectively. Compared to the voltage profile of case 1 in Figure 9.4c, the voltage profile obtained in this case is improved.

The ESS/BESS SOC curve and its Discharging and Charging power over a 24hr period is shown in Figure 9.7d. The Capacitor size and its cost at each hour is shown in Figure 9.7e.

For Case 4, the Active and Reactive Power LMP's at different bus for 24 hrs. period are plotted in Figure 9.7f. Here, the observation is that, LMPs of active power at peak load, i.e., at 18th hr. there is change in LMPs compared to the case 3 LMPs. At off peak load i.e., at 5th hr. the active power LMPs are almost equal.

For Case 4, Generator Economics calculations are calculated and the results are plotted in Figure 9.7g. In Figure 9.7g, at 5th hr the generator profit is positive but compared to that of in case 1, 2 and 3, also the noticeable

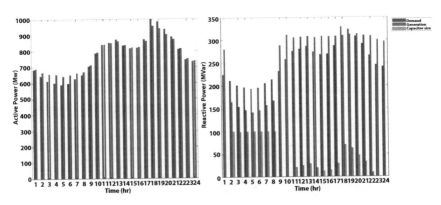

Figure 9.7a Case 4: Active and reactive power demand and generation and capacitor size.

Impact of BESS and DRP on LMP in Distribution System   267

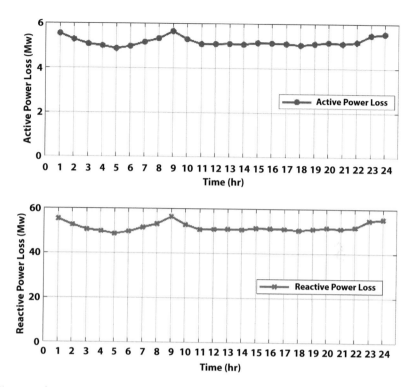

Figure 9.7b Case 4: Active and reactive power loss.

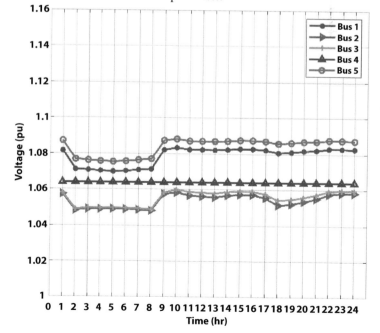

Figure 9.7c Case 4: Voltage profile.

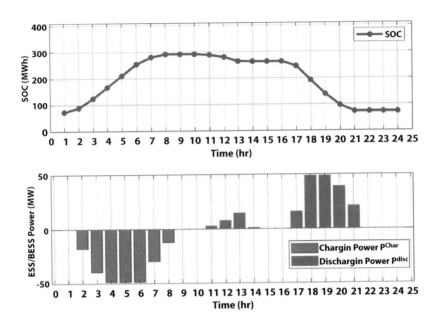

Figure 9.7d Case 4: ESS/BESS: SOC, discharging and charging power.

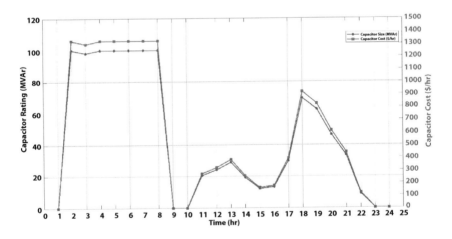

Figure 9.7e Case 4: Capacitor size and cost vs time.

point is that the cost of generation is less than the generator revenue and it can be witnessed in the zoom plot.

The important findings of the Case 4 results are that the LMPs of active power undergoes changes due to ESS/BESS installation and the voltage profile is improved by shunt capacitor.

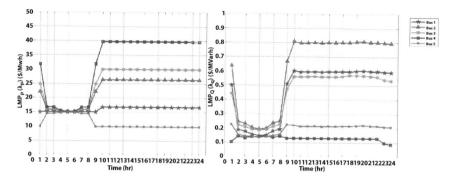

Figure 9.7f Case 4: LMPs of active and reactive power at different buses.

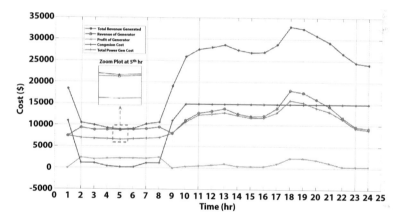

Figure 9.7g Case 4: Generator economics: cost and profit.

9.4.5 Case 5: Minimization of the Active Power Production and Reactive Power Production Cost

This case is an extension to case 4, the objective function and Shunt capacitor and ESS/BESS (installed at bus 4) ratings are the same. In this case, at bus 2 the load is assumed to be participated in Demand Response program and agreed to curtail 50 MW active power load. The Active and Reactive power Demand (Without DR) and Generation and the Capacitor size is shown in Figure 9.8a.

Due to the load participation in DR, the Demand at bus 2 is curtailed optimally and the new total demand is shown in Figure 9.8b and is compared with the demand without DR participation. There is a considerable variations in the demand during peak hours 18 to 21.

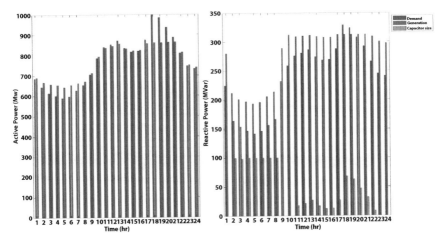

Figure 9.8a Case 5: Active and reactive power demand and generation and capacitor size.

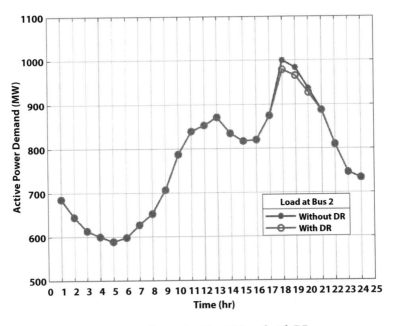

Figure 9.8b Case 5: Active power demand: without DR and with DR.

Impact of BESS and DRP on LMP in Distribution System

The active power curtailed power at bus 2 and the incentives provided for the curtailment is shown in Figure 9.8c.

From Figure 9.8c, it is seen that at 18th, 19th, 20th hr and 21st hr the active power load curtailed are 21.335 MW. 18.3173 MW, 9.6373 MW and 0.7107 MW, respectively. For the respective load curtailment, the incentives paid are \$465.837, \$344.68, \$97.6967 and \$0.86049 respectively. For the incentive payment, the previous case, i.e., case 4 active power LMPs at bus 2 are considered.

For Case 5, the hourly active and reactive power loss and the voltage profile at different bus are plotted in Figure 9.8d and Figure 9.8e, respectively.

The ESS/BESS SOC curve and its Discharging and Charging power over 24hr period for case 5 is shown in Figure 9.8f. The Capacitor size and its cost at each hour is shown in Figure 9.8g.

For Case 5, the Active and Reactive Power LMP's at different bus for 24 hrs. period are plotted in Figure 9.8h.

The LMP comparison of Active power of Case 4 and Case 5 at 18th, 19th, 20th and 21st hr is shown in Figure 9.8i. Due to the DR program the load is curtailed at hr where the peak of demand is observed; the loads which are participating in the DR program will now pay the price less than when

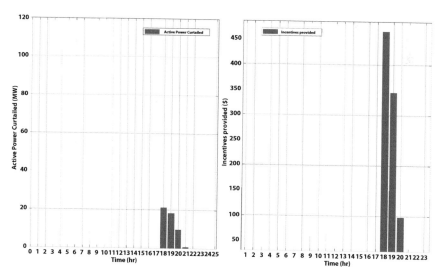

Figure 9.8c Case 5: Load curtailment and incentive payment.

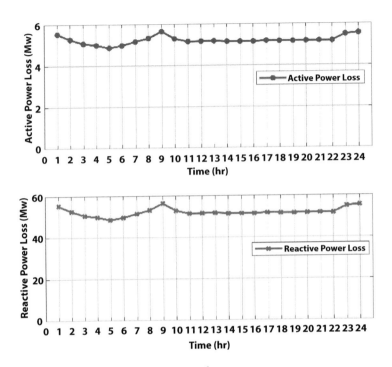

Figure 9.8d Case 5: Active and reactive power loss.

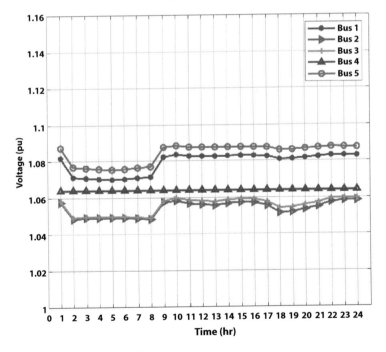

Figure 9.8e Case 5: voltage profile.

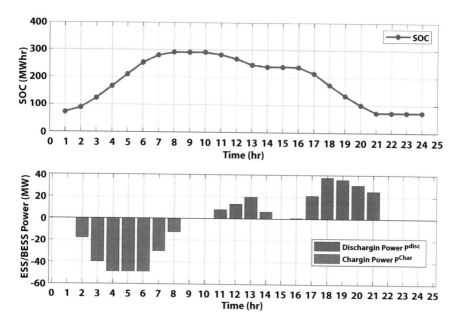

Figure 9.8f Case 5: ESS/BESS: SOC, discharging and charging power.

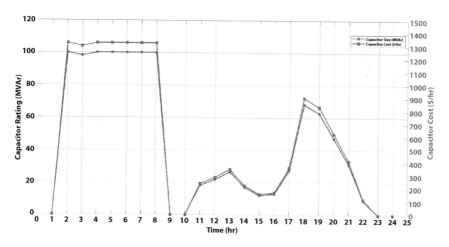

Figure 9.8g Case 5: capacitor size and cost vs. time.

they did not participate in the DR program. Here we compare the results of Case 4 and Case 5 active power LMP and a fair reduction in LMP is observed with the respective to the amount of load curtailed For Case 5, Generator Economics calculations are calculated and the results are plotted in Figure 9.8j. In Figure 9.8j, at 5th hr the generator profit is positive,

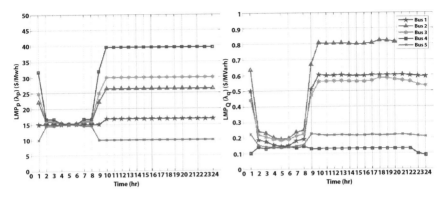

Figure 9.8h Case 5: LMPs of active and reactive power at different buses.

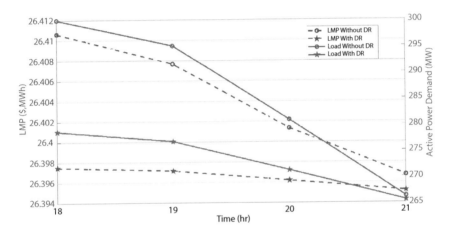

Figure 9.8i Case 5: comparison of active power LMP of case 4 and case 5.

also the cost of generation is less than the generator revenue and it can be witnessed in the zoom plot.

To summarize and compare the results of all 5 cases, for simplicity we only considered the two prominent points, i.e., the two hours i.e., 5th hr (Off-peak) and 18th hr (Peak) to observe the trends of results and the various parameters and are calculated and tabulated in Table 9.5 and Table 9.6 respectively. The active and reactive power LMPs at Off-peak and Peak hours are also tabulated in Table 9.5 and Table 9.6.

From Table 9.5, it is seen that the reactive power production cost is decreased in case 3, case 4 and case 5 while it is compared with the case 2. The active power production cost increases significantly from case 3 to case 5 due to the ESS/BESS installation, i.e., at this off-peak load period

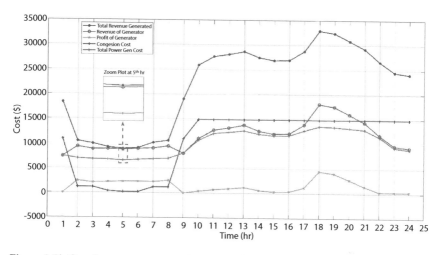

Figure 9.8j Case 5: generator economics: cost and profit.

Table 9.5 Result comparison of all cases at off-peak load.

Parameter		At off-peak load (at 5th hr)				
		Case 1	Case 2	Case 3	Case 4	Case 5
Active Power (MW)	Demand	588.8740				
	Generation	593.209	593.378	593.340	642.341	653.325
	Loss	4.335	4.504	4.466	4.860	4.987
Reactive Power (Mvar)	Demand	193.557				
	Generation	236.904	238.592	138.213	142.156	147.000
	Loss	43.347	45.035	44.656	48.599	49.874
Reactive Power Support by Shunt Capacitor (Mvar)		-	-	100	100	100
ESS/BESS	Charging Power (MW)	-	-	-	48.607	48.604
	Discharging Power (MW)	-	-	-	0.000	0.000

(*Continued*)

Table 9.5 Result comparison of all cases at off-peak load. (*Continued*)

Parameter		At off-peak load (at 5th hr)				
		Case 1	Case 2	Case 3	Case 4	Case 5
Active Power Curtailed (MW)		-	-	-	-	0
Incentives paid ($)		-	-	-	-	0
Active Power LMP's ($/MWh)	Bus 1	10.042	10.051	10.048	15.000	15.000
	Bus 2	10.163	10.204	10.188	15.209	15.209
	Bus 3	10.113	10.219	10.202	15.232	15.232
	Bus 4	10.108	10.140	10.129	15.136	15.136
	Bus 5	10.000	10.000	10.000	14.934	14.934
Reactive Power LMP's ($/MVArh)	Bus 1	0	0.252	0.144	0.148	0.148
	Bus 2	0.01029	0.290	0.176	0.193	0.193
	Bus 3	0	0.286	0.172	0.187	0.187
	Bus 4	0	0.268	0.140	0.141	0.141
	Bus 5	0	0.241	0.137	0.138	0.138
Active power production cost ($)		5932.087	5933.776	5933.396	6595.113	6759.868
Reactive power production cost ($)		-	29.954	10.014	10.668	11.425
Total power production cost (or)		5932.087	5963.730	5943.411	6605.781	6771.293
Shunt capacitor return/cost ($)		-	-	1324.000	1324.000	1324.000
BESS Cost ($ per day) (Operation + Investment)		-	-	-	2146.614	2146.504
Revenue Generated ($)		5975.600	6050.750	6019.113	8975.942	9258.868
Congestion Cost ($)		86.592	108.212	99.767	156.334	337.107
Revenue of Generator ($)		5889.008	5942.538	5919.346	8819.608	8921.760
Profit of Generator ($)		-43.079	-21.192	-20.984	2213.828	2150.467

Table 9.6 Result comparison for all the cases at peak load.

Parameter			At peak load (at 18th hr)				
			Case 1	Case 2	Case 3	Case 4	Case 5
Active Power (MW)	Demand		1000.000				
	Generation		1005.238	1004.645	1004.510	956.431	860.710
	Loss		5.238	4.645	4.510	5.038	5.157
Reactive Power (Mvar)	Demand		328.690				
	Generation		381.075	375.138	281.818	309.599	312.036
	Loss		52.385	46.448	45.097	50.377	51.566
Reactive Power Support by Shunt Capacitor (Mvar)			-	-	91.969	69.468	68.220
ESS/BESS	Charging Power (MW)		-	-	-	0.000	0.000
	Discharging Power (MW)		-	-	-	48.607	37.773
Active Power Curtailed (MW)			-	-	-	-	21.335
Incentives paid ($)			-	-	-	-	465.837
Active Power LMP's ($/MWh)		Bus 1	16.876	16.792	16.812	16.783	16.771
		Bus 2	26.503	26.416	26.422	26.411	26.398
		Bus 3	30.000	30.000	30.000	30.000	30.000
		Bus 4	39.684	39.862	39.781	39.730	39.644
		Bus 5	10.000	10.000	10.000	10.000	10.000
Reactive Power LMP's ($/MVArh)		Bus 1	0.347	0.658	0.556	0.598	0.602
		Bus 2	0.375	0.867	0.753	0.801	0.608
		Bus 3	0.122	0.680	0.560	0.571	0.584
		Bus 4	0	0.306	0.132	0.132	0.132
		Bus 5	0	0.276	0.180	0.217	0.216
Active power production cost ($)			17546.740	17640.889	17653.938	15717.607	13525.164

(*Continued*)

Table 9.6 Result comparison for all the cases at peak load. (*Continued*)

Parameter	At peak load (at 18th hr)				
	Case 1	Case 2	Case 3	Case 4	Case 5
Reactive power production cost ($)	-	89.295	56.171	64.194	65.659
Total power production cost (or)	17546.740	17730.184	17710.108	15781.801	13590.823
Shunt capacitor return/cost ($)	-	-	1217.672	919.754	903.235
BESS Cost ($ per day) (Operation + Investment)	-	-	-	2146.614	2092.350
Revenue Generated ($)	17901.076	33062.380	32986.057	32949.374	32933.084
Congestion Cost ($)	15044.979	14922.951	14871.921	14854.723	14844.774
Revenue of Generator ($)	17891.763	18139.428	18114.136	18094.651	18088.310
Profit of Generator ($)	354.337	409.245	404.027	2312.850	4497.487

the ESS/BESS undergoes charging with zero load reduction. The Active power LMPs from case 1 to case 3 increases (see in Table 9.5) and it's due to the charging of energy devices; also the reactive power LMPs from case 1 to case 5 decreases as the reactive power support is provided by shunt capacitor. It is also observed that the profit of generator is negative, i.e., generator has to face loss from case 1 to case 3, whereas from case 4 to case 5 the profit is positive and it is appreciated. The reason for the negative profit of generator is the cost of production is greater than the revenue of the generator, and it can be made positive by adding active power support devices like ESS/BESS, Renewable energy sources, etc.

From Table 9.6, the active power generation cost is reduced significantly from Case 3 to Case 5 along with the active power generation, due to the external active power support by ESS/BESS and load reduction through DR program. The reactive power generation cost also decreased from Case 2 to Case 5 as the shunt capacitor provided the reactive power support, thereby the requirement of reactive power generation also decreased. At peak load period, the ESS/BESS will start discharging the power and also the Load undergoes some curtailment as if it is willing to reduce by participation in the DR program, Load are encouraged to participate in DR program mainly at peak load period by providing them incentives.

The active power LMPs at bus 3 and 5 remains constant for all cases; however, at bus 2 and 4 it decreased from case 1 to case 2, and at bus 2 it increased from case 4 to 5. During peak load period, the generator profit is always positive irrespective of case and the profit increases from case 1 to case 5.

9.5 Conclusions

In this chapter, change in Locational Marginal Price (LMP) of active and reactive power are analyzed for different type of objective functions. The impact of installation of Shunt capacitor, Energy Storage System (ESS) like Battery Energy Storage System (BESS) and Load participation in Demand Response (DR) program on the LMPs is discussed. Technical aspects like losses, voltage profile are studied. The various economic factors, like revenue generated, congestion cost, generator revenue, and generator profit are studied. Based on the analysis carried out, the following conclusions are made from the case studies:

- From case 1, the active LMPs are Positive and the LMPs of reactive power are Positive and Negative. By this we can say the LMP can be positive or negative. Negative LMP at a bus means the penalty for injective the power and compensation for drawing power. Here, Losses occur at the generator during off peak load hour.
- From case 2, both active and reactive power LMPs are positive. Here, the voltage profile is improved compared with the case 1 and the losses are also reduced. The generator still faces loss at off-peak load.
- From case 3, due to the shunt capacitor installation the reactive power LMPs at bus 4 is reduced and also the voltage profile is improved. Here, we can say that the reactive power support devices installation will leads to the changes in the reactive power LMPs, i.e., either positive or negative based on the devices which drawing or injecting the reactive power.
- From case 4, due to the BESS installation at bus 4, the active power LMPs changes. Here, the LMPs of this case are compared with LMPs of case 3. During the charging period of BESS, the active power LMP increases because during this period the BESS act like an additional load. During the discharging period of BESS, the active power LMP decreases

because during this period the BESS will act like a source. Here in this case during off-peak load condition the generator profit is positive i.e., the revenue of the generator is more than the cost of production.
- From case 5, due to the load at bus 2 participation in DR program the decrease in active power LMP is observed. So, when the load is reduced the LMP also reduced.

So, when there is a change in active and reactive power there will be a change in the active and reactive LMPs. Here, in this chapter, the Shunt capacitor, Battery energy storage system and DR programs are considered, but Renewable energy sources can also be considered and their impact on LMP can also be observed for planning of the hybrid DERs integrated systems by the distribution system operator.

References

1. D. Nurse, "What is Locational Marginal Pricing (LMP)?," 11 Apr. 2018. [Online]. Available: https://www.energyacuity.com/blog/what-is-locational-marginal-pricing-lmp/#:~:text=LMPs%20are%20made%20up%20of,operating%20day%20to%20avoid%20volatility.
2. H. Liu, L. Tesfatsion and A. A. Chowdhury, "Locational marginal pricing basics for restructured wholesale power markets," in *2009 IEEE Power & Energy Society General Meeting*, 2009.
3. Fred C. Schweppe, Michael C. Caramanis, Richard D. Tabors, Roger E. Bohn, *Spot Pricing of Electricity*, Springer US, 1988.
4. M. L. B. a. S. N. Siddiqi, "Real-time pricing of reactive power: theory and case study results," in IEEE Transactions on Power Systems," *IEEE Transactions on Power Systems*, vol. 6, pp. 23-29, Feb. 1991.
5. Y. Dai, Y. X. Ni, F. S. Wen and Z. X. Han, "Analysis of reactive power pricing under deregulation," in *2000 Power Engineering Society Summer Meeting (Cat. No.00CH37134)*, 2000.
6. B. Mozafari, A. M. Ranjbar, A. R. Shirani and A. Mozafari, "Reactive power management in a deregulated power system with considering voltage stability: Particle Swarm optimisation approach," in *CIRED 2005 - 18th International Conference and Exhibition on Electricity Distribution*, 2005.
7. S. &. G. R. &. J. H. Hasanpour, "A new approach for cost allocation and reactive power pricing in a deregulated environment," *Electr Eng 91*, pp. 27-34, 2009.

8. A. Saranya, K. S. Swamp, "Evaluation of locational marginal pricing of electricity under peak and off-peak load conditions," in *2016 National Power Systems Conference (NPSC)*, 2016.
9. Soroudi A, *Power System Optimization Modeling in GAMS*, Germany: Springer International Publishing, 2017.
10. M. F. a. F. L. Alvarado, "Using utility information to calibrate customer demand management behavior models," in *2002 IEEE Power Engineering Society Winter Meeting. Conference Proceedings (Cat. No.02CH37309)*.
11. M. Fahrioglu and F. L. Alvarado, "Designing incentive compatible contracts for effective demand management," *IEEE Transactions on Power Systems*, vol. 15, pp. 1255-1260, 2000.
12. *Handbook on Battery Energy Storage System*, Philippines: Asian Development Bank, 2018.
13. U.S Dept. of Energy, "Benefits of Demand Response in Electricity Markets and Recommendations for Achieving Them: A Report to the United States Congress," February 2006.
14. Yong Fu and Zuyi Li, "Different models and properties on LMP calculations," in *2006 IEEE Power Engineering Society General Meeting*, 2006.
15. "GAMS Documentation Center," 11 November 2021. [Online]. Available: https://www.gams.com/latest/docs/index.html.
16. Brooke A., Kendrick D., Meeraus A., *GAMS: A User's Guide*, Redwood City: The Scientific Press, 1998.
17. E.Rosenthal, Sichard, *GAMS-A User's Guide*, Washington, DC, USA: GAMS Development Corporation, 2007.
18. Yun Dai and Yixin Ni and C. M. Shen and Fushuan Wen and Z. X. Han and Felix F. Wu, "A study of reactive power marginal price in electricity market," *Electric Power Systems Research*, vol. 57, pp. 41-48, 2001.
19. L. Zhou, Y. Zhang, X. Lin, C. Li, Z. Cai and P. Yang, "Optimal Sizing of PV and BESS for a Smart Household Considering Different Price Mechanisms," *IEEE Access*, vol. 6, pp. 41050-41059, 2018.
20. R. G. S. S. a. J. L. M. Moghimi, "Battery energy storage cost and capacity optimization for university research center," in *2018 IEEE/IAS 54th Industrial and Commercial Power Systems Technical Conference (I&CPS)*, 2018.
21. Fangxing Li and Rui Bo, "Small test systems for power system economic studies," *IEEE PES General Meeting*, 2010.

10

Cost-Benefit Analysis with Optimal DG Allocation and Energy Storage System Incorporating Demand Response Technique

Rohit Kandpal[1], Ashwani Kumar[1*], Sandeep Dhundhara[2] and Yajvender Pal Verma[3]

[1]*Department of Electrical Engineering, National Institute of Technology Kurukshetra, Kurukshetra, India*
[2]*Department of Basic Engineering, COAE&T, CCS Haryana Agricultural University, Hisar, India*
[3]*Department of Electrical and Electronics Engineering, UIET, Panjab University, Chandigarh, India*

Abstract

Economics plays a crucial role in fulfilling the sustainable development goals in the current world scenario. The enhanced competitiveness in the current power sector market due to deregulation makes reliable power distribution more of a techno-economic problem than a technical issue. The power loss minimization is a pressing issue while dealing with the distribution network as it is the lion's share of the total losses occurring in the system. In this study, an analysis is conducted on IEEE 33 bus radial distribution system (RDS) using grey wolf optimization for allocation of photovoltaic panel-based DG (PVDG), wind turbine-based DG (WTDG), diesel DG and energy storage system (ESS) for minimizing the energy losses. Further, the electric vehicle charging stations (EVCS) are optimally allocated along with the real-time pricing demand response technique, and its effects on distribution locational marginal pricing are evaluated. Moreover, the cost-benefit analysis of incorporating the demand response and increasing the number of energy storage systems is studied. The simulated results obtained are promising and make the distribution system more efficient.

---

*Corresponding author: ashwani.k.sharma@nitkkr.ac.in

Sandeep Dhundhara, Yajvender Pal Verma, and Ashwani Kumar (eds.) *Energy Storage Technologies in Grid Modernization*, (283–316) © 2023 Scrivener Publishing LLC

Keywords: Demand response, distribution locational marginal pricing, DG allocation, grey wolf optimization, energy storage system

10.1 Introduction

In the current scenario, countries are expected to adhere to the sustainable development goals adopted by the United Nations which are a blueprint for the peace and prosperity of people and the planet both in the present and future. One of the most significant of the sustainable development goals is averting the energy crisis arising due to the population explosion and industrialization. Power loss minimization is among the essential issues while adhering to the sustainable goal. As per studies, the distribution side is estimated to account for around 70% of overall power losses in the power grid, contributing a significant chunk of the total despite its low voltage operations, thus needing proper remedial measures. Deregulation of the power system with the enactment of the Electricity Act of 2003 brought about significant changes in the power sector with the influx of private players into the market. Thus, the economic aspect of the power distribution also became an immensely important feature in the power sector. Further, increased public awareness of environmental impacts along with the legal and international commitments, namely COP26 and the Paris Agreement, forced the government to incentivize renewable generation. Due to power deregulation, environmental concerns, and the depletion of fossil fuels, DG units have been widely linked to the distribution system in recent years. The unidirectional power flow of the traditional power value chain is replaced by the bidirectional flow with the integration of distributed generation. The rising power demands are thus curbed by accommodating the distributed generating sources such as micro-turbines, fuel cells, mini-hydro, battery storage, solar, and wind as nearer to the load as possible, reducing the burden on utility and in turn the power losses. The infiltration of renewable energy resources as distribution generation is increasing, thus reducing the dependency on centralized generation to a greater extent. The integration of the DGs may yield significant technical and economic benefits by feeding loads during peak load periods when energy costs are comparatively higher. However, technical and economic variables impact the penetration and practicability of DG in a particular region. The linking of DG to distribution networks adjacent to the load center potentially influences the direction and magnitude of network power flow, thus affecting the technical and economic bearing of distribution companies' network operations and planning procedures, whereas the inappropriate allocation of DG units will overshadow all aforementioned advantages. Index-based approaches [1–4], analytical-based [5, 6], and optimization-based methods all have been considered when determining the optimal allocation of DG [7, 8]. Various optimization methodologies have already been utilized to achieve diverse distribution system objectives. Genetic Algorithm [9], Bird Swarm Algorithm [10], Bat Algorithm [11], Grasshopper Optimization Algorithm [12], Particle Swarm Optimization [13–15], Bat Algorithm [16] Whale Optimization [17], Bald Eagle Search [18], Water Cycle Algorithm [19], Gravitational Search Algorithm [20] and Grey Wolf Optimization [21] have recently been used to calculate the best allocation of DG.

In this chapter, the optimal allocation of renewable DGs alongside thermal DG and energy storage system are evaluated on IEEE 33 bus radial distribution system. The EVCS is allocated at optimal positions considering not only the technical aspect but also the area served by it during its allocation. The analysis is carried out by modeling the load to make it more proximate to the real-time conditions. In addition, the chapter evaluates the cost-benefit analysis of increasing the number of energy storage units in the optimally allocated setup. Moreover, the effect of demand response techniques on the distribution location marginal pricing is evaluated and the cost-benefit analysis is obtained; incorporating the amount of energy storage units and the impact of the demand response technique is also studied. Moreover, the effect on the voltage profile of the system is observed with the implementation of demand response techniques and optimal allocation of DG power injection.

Grey wolf optimization is utilized for solving an optimization problem due to its prominent advantages over other similar optimization techniques because of its simple structure. Due to its structural advantage, the storage space requirement is reduced and with only two adjusting parameters it is simple and quick along with a fast convergence rate because of the diminished search space and fewer decisions. The results are obtained for an IEEE 33 bus distribution test system.

10.2 Distribution Generation and Energy Storage System

Distributed generation has been prominent in recent times but there is still some ambiguity surrounding the precise definition of DG. Although many scholars and organizations define it differently, it may be classified broadly

Table 10.1 Types of DGs and their performance parameters.

Distribution generation	Availability	Stability	Voltage control	Response speed
Biomass	Good	Good	Yes	Fast
Geothermal	Good	Good	Yes	Medium
Pumped Hydro	Good	Good	Yes	Fast
Solar Thermal	Uncertain	Poor	Uncertain	Variable
Photovoltaics	Uncertain	Poor	Uncertain	Fast
Wind	Uncertain	Poor	Uncertain	Fast
Low-head hydro	Variable	Good	Yes	Fast

based on size, position, voltage levels, penetration levels, fuel type, etc. Both renewables and non-renewables can be utilized as DG units. Most commonly used DG systems are PVDG, WTDG, fuel cells, natural-gas-fired REs, and diesel or gasoline-fuelled emergency backup generators. Sources like solar, wind, hydro, tidal, and geothermal are some of the renewable sources from which energy can be tapped, but among these solar and wind are the world's fastest growing because of ease of availability. Table 10.1 shows various kinds of DG used for various distribution sectors along with the performance factors.

The increasing congestion in the existing power system coupled with the lack of land availability for traditional power value chain construction has become an issue of concern for power sector companies. In addition to this, the increasing cost of fuels in the backdrop of liberalization and privatization of the electric market altogether raised the need for DG integration. With this background the DG technologies have emerged as reliable and viable alternatives for the customers and further due to the governmental support in the form of subsidies and policies towards the integration of renewable energy sources as DG, making them even more popular among the masses.

10.2.1 Renewable Energy in India

India's geographical location is a godsend for the utilization of renewable energy, with 300 sunny days and a coastline of 7500 km, both of which benefit the solar and wind industries, respectively. Additionally, the Indian

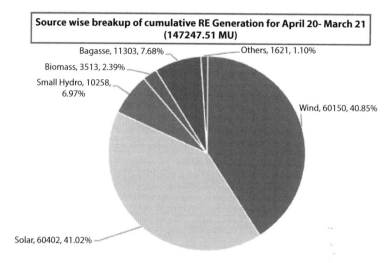

Figure 10.1 Indian renewable energy sector. [Source: Ministry of New and Renewable Energy.]

government's backing of renewable energy sources integration owing to its obligation towards international commitments like achieving net-zero emission by 2070 as proclaimed at COP26 further benefitted the industries. India is well on track to fulfilling these commitments, having increased its renewable energy capacity by two-thirds between April 2014 and January 2021. Globally, India today stands as 4th renewable energy power capacity, 4th in wind power and 5th in solar power capacity. Figure 10.1 shows the current renewable energy capacity of India. The Indian government's commitment to renewables is reflected in policies such as the Jawaharlal Nehru National Solar Mission and the Faster Adoption and Manufacturing of (Hybrid &) Electric Vehicles in India (FAME) [22]. These schemes will help decrease thermal generation, reducing the burden on foreign reserves whose significant share is the purchase of fossil fuels, while simultaneously creating a massive amount of sustainable jobs for the nation's youth.

10.2.2 Different Types of Energy Storage and their Opportunities

The present power system network is transforming the grid from the conventional to the smart grid with the amalgamation of conventional sources of energy and renewable energy sources (RES). With the intermittency of the RES, new options of energy storage technology are emerging to mitigate floating solar and wind power. The storage technologies are undergoing innovation to improve the performance of the grids.

A wide variety of different types of energy storage options are available for commercial use in the energy sector. These are emerging as the technology component in the energy systems to meet growing energy demand and balance the load demand mismatch. With the need for growth of energy storage in the electricity sector, so many technical challenges are emerging that require a range of solutions with the coordination of different storage devices and demand response program.

Utilities are looking for different ways and means to store energy that is produced at low load hours which can be used later to reduce imbalances between energy demand and energy production. Applications of energy storage devices are becoming more widespread and diverse with the growth of variable wind and solar power. With the output of wind and solar dependent on the local weather, variability can occur on different timescales; energy storage systems can provide fast responsive actions meeting variations in supply and demand.

The different types of energy storage systems can be categorized as:

1. Battery Energy Storage
2. Thermal Storage
3. Mechanical Storage
4. Pumped hydro Storage
5. Hydrogen Storage

These different storage devices have applications in different sectors like electricity, transport, heating of buildings, etc.

- **Battery storage systems:** Batteries are based on electrochemical technology that are comprised of a combination of cells. It is one of the oldest and most widely used forms of storage. Batteries may be of different types such as lithium-ion, and lead acid, and others can be of solid battery types like nickel-cadmium and sodium-Sulphur. Other types in the battery storage category contain liquid electrolyte solutions, such as vanadium redox and iron-chromium and zinc-bromine solutions. Super-capacitors are based on electrochemical technology that has a variety of applications in electrical utility but are not batteries.
- **Thermal storage:** Thermal storage devices are based on the physical capability of the absorption and release of heat or

cold in a form of solid, liquid, or air that involves changes in the physical state of the storage medium, which can be a gas to liquid or solid to liquid, and reverse. Thermal storage technologies include energy storage with phase change materials that can be molten salt and liquid air or cryogenic storage having the capability of storing heat/cold for a longer duration. Molten salt as a thermal storage system has emerged as the commercially viable solution with concentrated solar power. Phase change materials are playing a key role in storage technology.
- **Mechanical storage:** Mechanical storage systems are the simplest. They utilize the kinetic energy of rotation/gravitational energy to store energy in the form of mechanical energy as storage. The main options for mechanical storage are energy storage with flywheels and compressed air systems.
- **Pumped hydro storage:** Pumped hydro systems based on large water reservoirs are one of the best solutions offered as an energy storage system and have been one of the solutions used by the utility to meet load demand during peak hours. Such storage systems require water cycling between two reservoirs placed at different levels that store the energy in the potential form of energy in the reservoir located at a higher level. The power is generated when water stored in the reservoir at a higher level is released to the lower reservoir. The pumped storage is one of the best options to meet the variability of wind and solar power and to meet the load impedances in a short time frame.
- **Hydrogen storage:** Hydrogen storage systems are an emerging energy storage technology in the future. It involves its conversion from electricity via electrolysis for storage in tanks. From there it can later undergo either re-electrification or supply to emerging applications such as transport, industry, or residential as a supplement or replacement of gas.

Each of the different energy storage technologies has an impact on the system operation and needs to be considered for analysis and implementation. Key issues that need to be assessed are the charge, discharge profiles, storage capacity/capability, and potential scalability of the different storage devices. In addition, the cost of the storage, and the expected lifetime in terms of cycling frequency before degradation are some of the key issues that also need to be addressed in a cost-benefit analysis.

10.2.3 Distributed Generation

DG installation severely impacts various aspects of the power system, notably line losses, voltage profile, degree of injected harmonic, short circuit current, and systems reliability and stability. However, the system's reliability and stability would deteriorate on the installation of DG units at inconvenient places and of varied sizes, thus distorting the voltage profile and enhancing system losses rather than enhancing reliability and sustaining system stability making the optimal allocation of DG an issue of great importance.

10.2.3.1 Solar Photovoltaic Panel-Based DG (PVDG)

The solar radiation for the examined region of Bhuj, Gujarat, is acquired to compute the output power of the unit PV module using eq (10.1) with the parameters specified in Table 10.2 [23, 24]. The power generation profile of PV system is displayed in Figure 10.2.

$$P_{unit-pv} = (P_{rated} A_m D_f) \left(\frac{G}{G_{ref}} \right) \left(1 + K_T (T_c - T_{ref}) \right) \quad (10.1)$$

Table 10.2 PVDG unit specification.

Parameters	Data
PV panel rating (P_{rated})	1 kW
Temp coefficient of power (K_T)	-0.44%
Derating factor (D_f)	0.88
Solar radiation incident (G)	0.9873 kW/m²
Standard incident radiation (G_{ref})	1 kW/m²
Current temp. (T_c)	33.83 °C
Reference temp. (T_{ref})	25°C

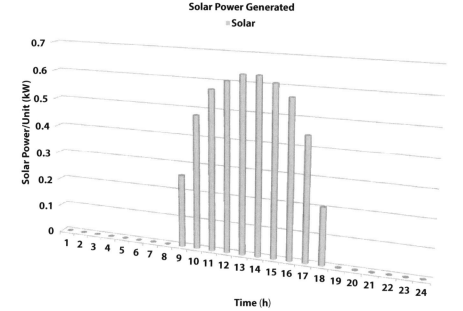

Figure 10.2 PVDG generation profile.

The total solar power generated can be calculated by using eq (10.2)

$$P_{PV} = P_{unit-pv} * N_{pv} \qquad (10.2)$$

10.2.3.2 Wind Turbine–Based DG (WTDG)

The speed of wind for the examined region of Bhuj, Gujarat, is acquired to compute the output power of the unit WT module using eq (10.3) with the parameters specified in Table 10.3 [23, 24].

$$P_{unit-wt} = \begin{cases} 0 & V_{out} \geq V(t) \leq V_{in} \\ P_T^w & V_r \leq V(t) \leq V_{out} \\ P_T^w \dfrac{V(t) - V_{in}}{V_{rat} - V_{in}} & V_{in} \leq V(t) \leq V_r \end{cases} \qquad (10.3)$$

Where, V(t) and V_r(t) are the wind speed at the desired & reference height, respectively, which is calculated by using eq (10.4) and total wind power generated using eq (10.5). The power generation profile of wind turbine generator is displayed in Figure 10.3.

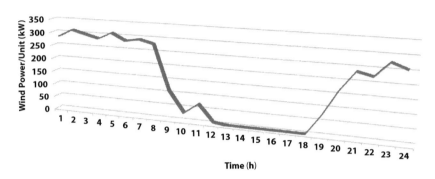

Figure 10.3 WTDG generation profile.

$$V(t) = V_r(t) \left(\frac{H_T}{H_{ref}} \right)^{\gamma} \quad (10.4)$$

$$P_W = P_{unit-wt} * N_{wt} \quad (10.5)$$

Table 10.3 WTDG unit specification.

Parameters	Data
Rating of wind turbine (P_{rated})	1500 kW
Rated speed of turbine (V_{rat})	14 m/s
Reference height (H_{ref})	10 m
Friction coefficient (γ)	1/7
Cut-in speed (V_{in})	3.5 m/s
Cut-out speed (V_{out})	20 m/s
Speed at reference height (V_r)	2.973 m/s
Height of shaft (H_T)	75

10.2.3.3 Load Model and Load Profile

An overwhelming bulk of load flow studies consider the loading in the system to be constant during the analysis but this assumption is far from reality. On the other hand, load variation is a significant factor in distribution system analysis. The loading of the system depends immensely on the operating conditions of the system; therefore, its effect needs to be considered for power deployment. The realistic distribution system incorporates all types of loading, namely residential, industrial and commercial, and the system's frequency and voltage play a pivotal role in the analysis. However, as the frequency of the system cannot be altered locally, acclimating load voltage will vary total power consumption according to the equations given below with the parameters for voltage dependency of the various loading stated in Table 10.4 [25].

$$P_L = P_{L_o}\left(\frac{V}{V_{ref}}\right)^\alpha$$

$$Q_L = Q_{L_o}\left(\frac{V}{V_{ref}}\right)^\beta \tag{10.6}$$

Each bus incorporates all three types of loads in modest amounts of the overall load at any stipulated time and the total load on any stipulated time is calculated using eq (10.7). The proportion for every type of load in total load varies with time, as seen in Figure 10.4 [26] along with the load factors for the day taken into account [27].

$$S_T = K_{res} \cdot S_{res} + K_{ind} \cdot S_{ind} + K_{com} \cdot S_{com} \tag{10.7}$$

Table 10.4 Load modeling parameters.

No.	Load types	α	β
1.	Industrial	0.18	6
2.	Commercial	1.25	3.5
3.	Residential	0.72	2.96

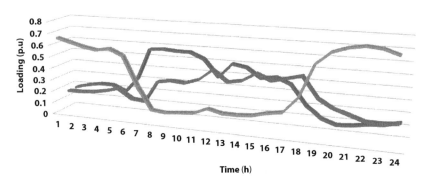

Figure 10.4 Load variation profile of a day.

10.2.4 Demand Response Program

Demand-Side management is planning, implementing, and evaluating activities formulated to encourage customers for modifying their energy consumption by implementing the three broad measures to elicit consumer response. Firstly, the consumers can mitigate their power usage during peak periods when tariffs are usually higher without disrupting their trends during off-peak periods. Secondly, consumers might adapt to high expenses by migrating certain peak demand consumption to an off-peak period. The last type of consumer response is the on-site generation of power by allocation of DG [28–30]. The demand-side management programs can be categorized into two categories, namely energy saving or conservation process and load management programs. Load management or rather commonly known as demand response programs are highly favored, thus further discussion will be based on demand response techniques. Demand response refers to variations in power usage by end-use customers from their customary consumption patterns as a response to changes in power prices across the board. The demand response techniques can be further classified as incentive-based programs and price-based programs [31]. Further segments of incentive-based programs are as follows:

- *Market-Based:* The market forces incentivize the consumers for their active participation by submitting offers for load reduction. Although large consumers can participate

directly there is provision for even small consumer participation via indirect means.
- **Direct Control–Based:** The utility directly controls a specific type of appliance with the right for interruptions owing to economic or reliability reasons in the end-user premises without any pre-notification and engaging the participation of a large number of consumers. They control the number and duration of interruptions without compromising the comfort level of the participating consumer while compensating them with benefits while billing.
- **Curtailable-Based:** They are basically for medium and large consumers who are being incentivized for curtailing their specific loads for a specific time. These curtailments are mandatory and failure to comply may lead to penalties.

On the other hand, the price-based programs can be mainly segmented into the following:

- **Time of Use (TOU):** Traditionally end users charged with flat prices depicting the average cost of supplying electricity were not considerate of the varying cost of electricity. The TOU pricing reflects the variation of cost with time with the variation to day or season. The use of stepped rate reflects the average market price during consumption rather than the day-to-day volatility of cost.
- **Critical Peak Pricing (CPP):** Though the long-term supply cost is addressed by TOU there is still a need to capture the short-term cost. They superimpose the variation in rates owing to system criteria on either TOU rates or flat rates. The contractual understanding fixes the maximum number of periods for the application of CPP rates. There are two variates of CPP, Extreme Day pricing (EDP) and Extreme Day CPP (ED-CPP). The first applies higher rates for the entire day in case of the extreme day while the latter uses peak and off-peak rates only for extreme days while the rest of the days are computed by flat rates.
- **Real-Time Pricing (RTP):** In this, the energy prices are updated at very short intervals. This charges the consumers according to the hourly fluctuating rates that reflect the cost of power in real-time scenarios in the wholesale market by noticing the consumers of the price changes on an hourly

or daily basis. Customers that engage in DR programs can expect to save finances on their power bills if they avert their usage during peak hours [29].

The impact of the demand response technique can be determined by a reduction in the cost of peak demand saving with the reduction in peak demand expressing the potential of the demand response strategy. The reduction of peak demand along with energy consumption enhances the capability of the distribution system for supplying additional loads without the need for additional investments. Reduction in peak demand consequently aids in a reduction in construction cost and investments along with lower power tariffs thus diminishing the average price of energy. Cost of saving also includes the costs avoided for expansion required to meet the demand without a demand response strategy in addition to some operating cost savings; on the other hand the cost incurred during implementation of a demand response strategy also varies. The demand response technique also helps address the power shortage issues along with efficient management of system resources over time, avoiding involuntary outages and reducing the heavy cost incurred during power interruptions.

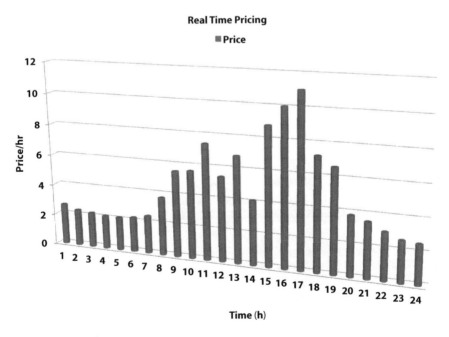

Figure 10.5 Real time pricing.

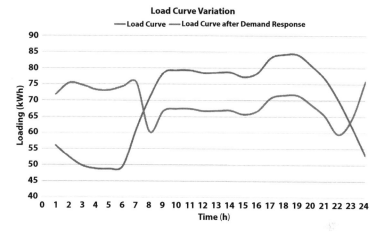

Figure 10.6 Variation in load curve due to RTP DR.

Various demand response techniques are utilized for gaining economic benefits but according to the beliefs of many economists RTP schemes are the most expedient DR programs viable for the competitive power sector, and governments should prioritize them. Due to better effective infrastructure utilization, overall power rates are destined to drop. On application of the RTP-DR technique, the peak-to-average ratio diminishes due to the shifting of load from the peak period to the off-peak period from 1.22 in the base case to 1.09 on the application of RTP-DR. The RTP scheme is indicated in Figure 10.5 [32] and the variation in the load profile on the application of the RTP-DR scheme is depicted in Figure 10.6.

10.2.5 Electric Vehicles

A transformation in the century-old automobile industry is imminent due to the spike in the price of fossil fuels. Adding to the economic issues the transportation sector contributes about a quarter of greenhouse gas (GHG) emissions, thus having a negative effect on the environment. As already discussed, India's commitments on international platforms to contain pollution and reduce carbon footprint forces the country to prepare itself for a shift toward EVs by 2030 [33]. These EVs are energy efficient, generating less GHG emissions and reduced noise. The EVs cut emissions and reduce the dependency on imported fuel, thus relieving India's FOREX against the vulnerability against crude prices and currency fluctuations. Although the capital required for EV is on the higher side but the low running and maintenance cost compensates for it. Low energy consumption and power tariffs result in lower running costs. EVs are propelled via electric motors along with a rechargeable battery or other portable energy storage device for maintaining the power supply. The different categories of EVs are namely HEV, PHEV, and BEV.

- *Hybrid electric vehicles (HEVs):* They are both fuel and electric powered with the battery system being charged by the power generated by the breaking system.
- *Plug-in hybrid electric vehicles (PHEVs):* They are similar to HEV but with a smaller engine and battery system with the charging done by either breaking system or by plugging into an external source.
- *Battery electric vehicles (BEV):* These depend on only the combination of battery and electric motor for propulsion with the charging dependent on external plugging points.

With the government's initiative for the reduction in petrol-driven vehicles, while giving a boost to electric vehicles due to various environmental, economic, and international relational reasons, there has been a significant investment in electric mobility. Electric vehicle charging stations (EVCSs) which are essential infrastructure for EVs, are now being erected with government and private sector backing. Accessibility to renewable DGs on a large scale and widespread usage of electric vehicles have boosted clean energy consumption and enhanced the long-term viability of power grid development. However, concerns like power quality reductions, voltage overruns, and a surge in system losses have emerged simultaneously. Their

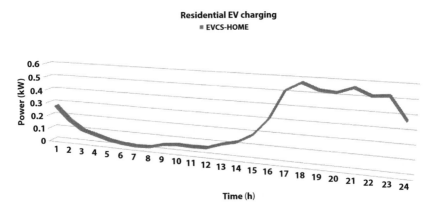

Figure 10.7 EV charging profile.

deployment, however, necessitates a thorough grasp of loading patterns; in order to truly comprehend the implications of EVs on the grid, EV load models are required. Figure 10.7 depicts the charging pattern of EVCS [34]. If the grid's balance is not adequately maintained, a substantial surge in the number of EVs in circulation leads to an increase in power usage, culminating in a blackout. Therefore, its allocation is of utmost importance, but inappropriate EVCS placement has a detrimental effect on the electric grid's efficiency. Incorporating EVCS into the electric distribution network enhances power loss thus degrading the voltage profile.

10.2.6 Modeling of Energy Storage System

With the colossal amount of RE integration into the system as discussed in the sections above, the need for peculiar flexible storage systems arises as it acts as a catalyst in India's race to fulfill its international commitments. Owing to the fluctuating nature of renewable energy sources there is a need for absorption and relinquishing the power as per the grid requirements to always keep up with the demand [35]. The challenge of grid balancing tends to become even more complex due to variable generation and ever-changing demand patterns causing an imbalance both daily and seasonally. While the seasonal imbalances are catered by the market forces the daily variations can be resolved by utilizing the energy storage system. The energy storage system is utilized for the storage of surplus energy during peak generating periods and discharging during off-peak periods. Thus, an energy storage system provides an economically viable option for the flexible and reliable system with the cost of energy storage systems declining sharply in the past decade [36]. Even though the energy storage system has the above-stated advantages it comes with a fair share of challenges attached to itself. The major one is having a grip on the supply chain required for its production, i.e., the availability of lithium, nickel and cobalt are a pressing issue. A thorough diagnosis of the state of charge (SOC) facilitates the measurement of energy by using eq (10.8–10.11) [37].

$$SOC(t+\Delta t) = SOC(t)(1-\phi) + \left(P_c^k \cdot \eta_c - P_d^k \cdot \frac{1}{\eta_d} \right) \cdot \Delta t \quad (10.8)$$

$$SOC_o = 0.85 \cdot SOC_{rated} \quad (10.9)$$

$$SOC_{max} = 0.99 \cdot SOC_{rated} \quad (10.10)$$

$$SOC_{min} < SOC(t) < SOC(t)_{max} \qquad (10.11)$$

10.2.7 Problem Formulation

In this chapter, the main objective is to minimize the energy loss in the distribution system. The minimization of the total power loss occurring throughout the day is presented as an objective function in (10.12).

$$Min(P_{Total}) = \sum_{t=1}^{24}(P_{loss}(t))\Delta t = \sum_{i=1}^{24}\left(\sum_{br=1}^{N} r_{br}^i \left(\frac{(P_{br}^i)^2 + (Q_{br}^i)^2}{(V_n^i)^2}\right)\right)\Delta t_i \qquad (10.12)$$

Where the number of branches represented by N and the P_{br}^i & Q_{br}^i representing the real and reactive power following in the branches respectively and V_n^i is the voltage at n^{th} bus at any time instance Δt subject to equality and inequality constraints, namely

- Voltage constraint: Retention of the distribution system's power quality.

$$|V_{min}| \leq |V_i| < |V_{max}| \qquad (10.13)$$

- Thermal limit constraint: The branch current's must be well within the conductor's maximum thermal capacity.

$$|I_m| \leq |I_{rated}| \qquad (10.14)$$

- Power penetration constraint: To prevent backflow, each embedded DG unit's total active power generation must be less than the network's total active power consumption.

$$0 \leq \sum_{i=1}^{n} P_{DG_i} \leq \sum_{i=1}^{n} P_{L_i} \qquad (10.15)$$

- Power Balance Constraints: The active power balance for k^{th} hour at i^{th} bus is represented as

$$P_i^k(Pg_i^k + Pgrid_i^k - Pd_{Load,i}^k)$$
$$= V_i^k \sum_{j=1}^{n} V_j^k (G_{ij}^k cos(\delta_i^k - \delta_j^k) + B_{ij}^k sin(\delta_i^k - \delta_j^k)) \quad (10.16)$$

Where Pg_i^k is the total active power generated from various sources incorporated as seen from the equation below

$$Pg_i^k = P_{dg,i}^k + N_{wind}(i) \cdot P_{wind_i}^k + N_{PV}(i) \cdot P_{PV_i}^k + N_{batt}(i)$$
$$\cdot \left(P_{ch_i}^k - P_{dis_i}^k\right) \quad (10.17)$$

Where $P_{dg,i}^k, P_{wind_i}^k, P_{PV_i}^k$, are the active power injected by diesel generators, WTDG, and PVDG. Whereas the $N_{wind}(i), N_{PV}(i), N_{batt}(i)$ represent the number of PVDG, WTDG, and ESS units, respectively alongside with $P_{ch_i}^k, P_{dis_i}^k$ being the charging and discharging power of ESS.

10.2.8 Distribution Location Marginal Pricing

Distribution Locational marginal pricing (DLMP) refers to the marginal cost of supplying the next incremental unit of load (MW) at a specific location. With the high penetration of distributed energy resources alongside the energy storage system and the varying demand of the distribution system, the optimal operation of the distribution system is a challenge, especially with the deregulation of the system having rekindled the interest of DLMP in the current scenario. The increased competitiveness due to this can be ensured by utilizing DLMP as it considers both the technical and economic concerns of the distribution system. Thus creating a chance of having third-party aggregators actively participating in the energy market by supporting their economic decision on sizing and placement of their own DGs. DLMP is adopted to mitigate the congestion on the distribution level associated with high penetration DGs. Thus the LMP depends on the location and size of DG along with the time interval. The real and reactive both power flows are considered for the computation of the DLMP [38]. The losses are computed as percentages and denoted by Kp and Kq.

$$C(P_{dg}) = a \cdot P_{dg}^2 + b \cdot P_{dg} + c \quad (10.18)$$

The Lagrange equation is utilized for the finally obtaining DLMP(λ)

$$C(P_{dg}) + \lambda(PD - \sum_{i=1}^{Ng} PGg + Kp * \sum_{i=1}^{Ng} PGg + Kq * \sum_{i=1}^{Ng} PGg) \quad (10.19)$$

$$\frac{\partial L}{\partial PGg} = \frac{\partial C(P_{dg})}{\partial PGg} - \lambda(1 - Kp - Kq) \quad (10.20)$$

$$\lambda = \frac{\frac{\partial C(P_{dg})}{\partial PGg}}{(1 - Kp - Kq)} \quad (10.21)$$

$$\lambda = \frac{2a \cdot P_{dg} + b}{(1 - Kp - Kq)} \quad (10.22)$$

10.3 Grey Wolf Optimization

Syedali Mirjalili et al. presented the "Grey Wolf Optimizer" (GWO) [39] in 2014, which is a population-based novel meta-heuristic optimization approach based on emulating the social hierarchy and hunting mechanism of a pack of grey wolves. The alphas, though not the strongest, are typically in charge of the decision-making aspects of the pack and their judgments are dictated and enforced throughout the pack by their subordinated betas. They are next in line and are most likely to take over from the existing alphas. The omegas are the least ranked, which makes them always subservient to the rest of the pack. The remaining are the deltas which are below the alphas and betas yet dominate the omegas. Thus, GWO mathematically emulates this hierarchical model of the pack of grey wolves by preserving the optimal solution to be alpha and subsequently making the beta and delta as the second and third best solutions, respectively. The hunting is steered by the alphas, betas, and deltas. Emulating the real-life hierarchy it is assumed that they are the three best solutions knowing the prospective prey's location. The wolves spread out in quest of prey and converge for encircling of prey and at last attack. To mathematically replicate the attack, a is diminished and consequently, the range of A shrinks as A=[-2a,2a] in which a diminishes from 2 to 0 throughout the course of iterations. To depict this behavior, we employ the variable A with random values with |A|>1 encouraging the grey wolves to diverge from the prey in the hopes of finding a fitter prey, whereas |A|<1 compels the wolves to strike the prey. Furthermore, the C comprises random values in the range [0,2], which endow the prey with random weights. This fosters more random behavior during optimization and aids in the dodging of local optima. C might even simulate the impediments that the wolves confront while approaching the prey in the physical realm, hindering the wolves from acquiring the prey directly and spontaneously. Thus, the social hierarchy, searching, and attacking the prey are all mathematically modeled using the following equations (10.23–10.33)

$$\vec{D} = |\vec{C}.\vec{X}_p(t) - \vec{X}(t)| \qquad (10.23)$$

$$\vec{X}(t+1) = \vec{X}_p(t) - \vec{A}.\vec{D} \qquad (10.24)$$

$$\vec{A} = 2\vec{a}.\vec{r}_1 - \vec{a} \qquad (10.25)$$

$$\vec{C} = 2.\vec{r}_2 \qquad (10.26)$$

Algorithm 10.1 Optimal allocation of Distribution Generation

Input: Bus Data, EV charging Data, Ratings (PV & WT), DR, Number of ESS
Output: Optimal allocation
Search agent initialization

1: Evaluate Load Flow
2: **while** Run GWO till Maximum iteration reached **do**
4:   Run respective cases
5:   Call LF to evaluate APL
6:   **if** (Power Loss well within Constrains) **then**
7:     Update the position of wolves
8:   **else**
9:     Relinquish Solution
10:   **end if**
11:   **if** (Solution obtained better than last run) **then**

12:     Supersede Solution
13:     else
14:         Reiterate GWO
15:     end if
16: end while
17: return *Optimal sizing & locations*

$$\vec{D}_\alpha = |\vec{C}_1.\vec{X}_\alpha - \vec{X}| \tag{10.27}$$

$$\vec{D}_\beta = |\vec{C}_2.\vec{X}_\beta - \vec{X}| \tag{10.28}$$

$$\vec{D}_\delta = |\vec{C}_3.\vec{X}_\delta - \vec{X}| \tag{10.29}$$

$$\vec{X}_1 = \vec{X}_\alpha - \vec{A}_1.(\vec{D}_\alpha) \tag{10.30}$$

$$\vec{X}_2 = \vec{X}_\beta - \vec{A}_2.(\vec{D}_\beta) \tag{10.31}$$

$$\vec{X}_3 = \vec{X}_\delta - \vec{A}_3.(\vec{D}_\delta) \tag{10.32}$$

$$\vec{X}(t+1) = \frac{\vec{X}_1 + \vec{X}_2 + \vec{X}_3}{3} \tag{10.33}$$

10.4 Numerical Simulation and Results

The simulation is performed on the most commonly used distribution test system, namely IEEE 33 bus test system. The total power requirement of the test case system is fed by the substation when no distribution generation is equipped. Along with the base loading the system is also integrated with EVCS charging stations at optimal locations thus increasing the loading compared to the test base system. The system time-varying modelling along with the EVCS is analyzed for optimal allocation of renewable DG allocation utilizing the grey wolf optimization and direct load flow [40] is carried out for calculation of various parameters. The analysis proposed the benefits of demand response and battery energy storage devices with the increase in the number of batteries are studied. The analysis is carried out for the four various cases considered as follows:

Case 1 (C1): Optimal PVDG, WTDG, Thermal DG, and 1 ESS without Demand Response.
Case 2 (C2): Optimal PVDG, WTDG, Thermal DG, and 1 ESS with Demand Response.
Case 3 (C3): Optimal PVDG, WTDG, Thermal DG, and 2 ESS without Demand Response.
Case 4 (C4): Optimal PVDG, WTDG, Thermal DG, and 2 ESS with Demand Response.

Table 10.5 shows the cost-benefit analysis of increasing the number of batteries and incorporating the RTP demand response technique. The cost of generation from the thermal DG which is allocated along with batteries and a combination of WTDG and PVDG alongside with the cost of the battery is incorporated. While keeping the renewable power generation the same for all the cases the cost of generation using the thermal DG varies in all cases. Considering C1 as the base case and considering the total combined cost of a single unit of battery to be 50000, the total cost-benefit analysis is considered for the subsequent cases. The cost-benefit analysis observed in the Table shows a benefit of 8.25% in C2 and a 39.79% benefit for C3. The highest benefit seen is of 50.89% in the case of C4.

The DLMP variation depends on the time, location, and sizing of the DGs. The location of the DG is kept constant in the analysis, thus DLMP is affected by the time and sizing of DG which can be collaborated as the operating conditions of power injection via DG throughout the day.

Table 10.5 Cost-benefit analysis of various cases.

	C1	C2	C3	C4
Cost of Generation DG ($)	1239753.6	1133319	676442.55	533407.7
Cost Battery ($)	50000	50000	100000	100000
Total Cost ($)	1289753.6	1183319	776442.55	633407.7
% Cost Benefit DG	-	8.25%	39.79%	50.89%

Figure 10.8 shows the variation in distributed locational marginal pricing on the implementation of RTP-DR with the variation closely following the load demand curve variation. Thus, it can be observed that the demand response technique has an impact on the location marginal pricing. Figure 10.9 shows the power generated by the thermal DG for all the cases. From the Figure, it can be observed that the increase in the number of the energy storage system and application of the demand response approach individually, as well as their combined implementation, immensely affect the requirement of power derived from the operation of thermal DG. With the total generation from thermal DG reduced by 8.58% on implementation of the demand response technique while by 45.44% on increasing the number of the energy storage device by one, whereas on combining the effect of both the parameters the total thermal DG generation requirement shows a reduction of 56.97%, thus aiding in the cost-saving with combined approach enhancing the saving to an even greater extent.

The voltage profile also shows an enhancement in the application of DG and further in the implementation of the demand response approach which can be seen in Figures 10.10–10.12 with the minimum voltage being 0.9332 p.u, 0.9595 p.u, and 0.9628 p.u for without DG, with DG, and with DG alongside with RTP-DR approach, respectively. Thus, the voltage profile is enhanced with the implementation of DG and further with the integration of demand response techniques. The optimal allocation of DG shows a 2.81% enhancement in the minimum voltage occurrence in

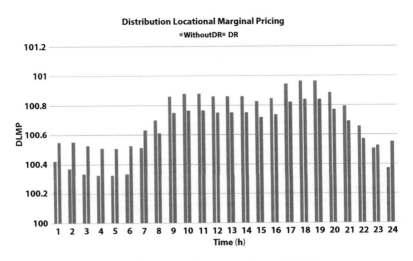

Figure 10.8 Variation in location marginal pricing owing to RTP-DR.

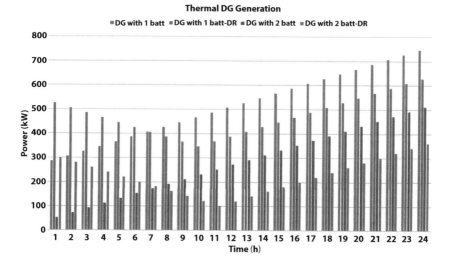

Figure 10.9 Variation in generation from thermal DG for various cases.

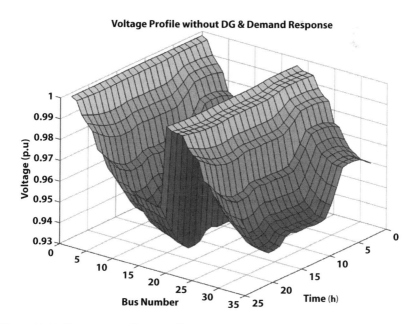

Figure 10.10 Variation in voltage profile without DG & demand response.

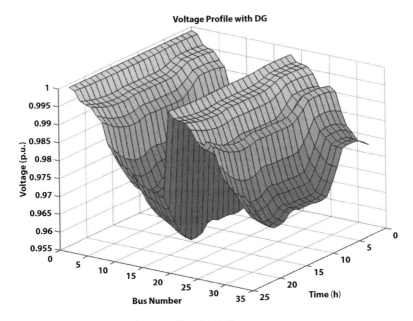

Figure 10.11 Variation in voltage profile with DG.

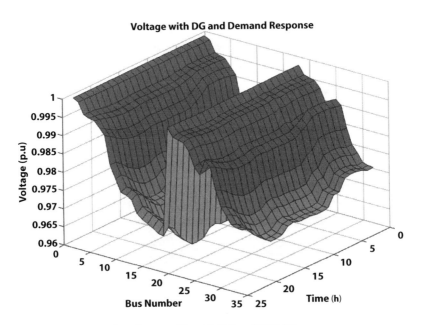

Figure 10.12 Variation in voltage profile with DG and RTP-DR.

the system throughout the entire day, whereas the integration of demand response alongside with DG further enhances the minimum voltage to 3.17% compared to that on the system without any optimal DG allocation or application of any demand response techniques. This improves the power quality of the system, which in turn will boost the profitability of the utility.

Figures 10.13–10.18 show the battery performances with each showing the SOC charging and discharging power of the energy storage system, for all the respective cases. The percentage SOC is depicted by the black line, whereas the red bars are for discharging power and the blue bars depict the charging power in the figures. The power supplied by the energy storage system is represented by negative power, whereas the power drawn in by the energy storage system is taken positive. The charging and discharging of energy storage systems depends on the availability of either excess or deficit of renewable energy systems. It can be observed that when utilizing only one energy storage system the battery reaches both extreme conditions which in turn adversely affects the battery life, whereas on increasing the number of energy storage units the battery performance enhances much compared to the case when it was operating alone, whereas the secondary battery has a much more stable operation which over time will lead to more cost benefit with the increased battery life.

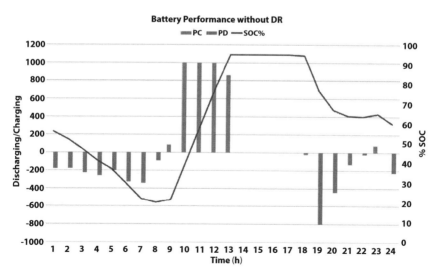

Figure 10.13 Battery performance of single battery without DR.

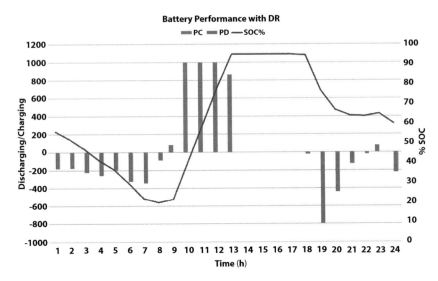

Figure 10.14 Battery performance of single battery with DR.

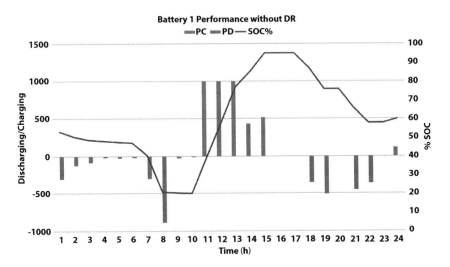

Figure 10.15 Battery performance of 1st battery without DR.

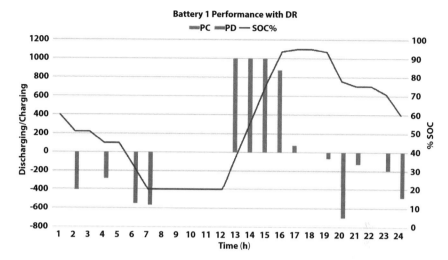

Figure 10.16 Battery performance of 1st battery with DR.

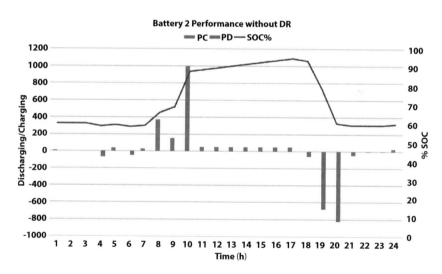

Figure 10.17 Battery performance of 2nd battery without DR.

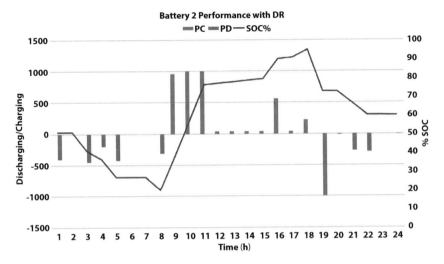

Figure 10.18 Battery performance of 2nd battery with DR.

10.5 Conclusions

The authors investigated the influence of RTP-DR approach and variation in the number of energy storage systems along with the optimally allocated DGs for the minimization of energy losses. It can be observed that the RTP-DR brings down the peak to average ratio, which helps to splay the load curve, thus shrinking the requisite actual power injection to provide for peak load. Thus, this variation in the load curve forces the variation in the operational requirement of the thermal DG unit, which in turn varies the distribution locational marginal pricing. The variation in locational marginal pricing can be utilized to enhance the techno-economic aspects of deployment of DG. Further from the cost-benefit analysis performed it is observed that the implementation of demand response provides a slight benefit but when combined with the increase in the number of energy storage system along with the demand response provides a significantly larger benefit in terms of cost incurred. This analysis will significantly aid the DNO in achieving a better distribution network, considering the RTP-DR approach and the number of energy storage systems needed to be deployed for the effective economic and technical operation for improved system quality and reliability to serve consumers with a better voltage profile and diminished loss reduction, resulting in increased network efficiency.

References

1. F. Iqbal, M. T. Khan, and A. S. Siddiqui, "Optimal placement of DG and DSTATCOM for loss reduction and voltage profile improvement," *Alexandria Engineering Journal*, vol. 57, no. 2, pp. 755–765, Jun. 2018, doi: 10.1016/j.aej.2017.03.002.
2. S. A. A. Kazmi, H. W. Ahmad, and D. R. Shin, "A New Improved Voltage Stability Assessment Index-centered Integrated Planning Approach for Multiple Asset Placement in Mesh Distribution Systems," *Energies (Basel)*, vol. 12, no. 16, Aug. 2019, doi: 10.3390/en12163163.
3. A. R. Gupta, "Effective Utilisation of Weakly Meshed Distribution Network with DG and D-STATCOM," *Journal of The Institution of Engineers (India): Series B*, vol. 102, no. 4, pp. 679–690, Aug. 2021, doi: 10.1007/s40031-021-00597-3.
4. J. S. Bhadoriya and A. R. Gupta, "A novel transient search optimization for optimal allocation of multiple distributed generator in the radial electrical distribution network," *International Journal of Emerging Electric Power Systems*, vol. 23, no. 1, pp. 23–45, 2022, doi: doi:10.1515/ijeeps-2021-0001.
5. S. Azad, M. M. Amiri, M. N. Heris, A. Mosallanejad, and M. T. Ameli, "A novel analytical approach for optimal placement and sizing of distributed generations in radial electrical energy distribution systems," *Sustainability (Switzerland)*, vol. 13, no. 18, Sep. 2021, doi: 10.3390/su131810224.
6. M. Aryanfar, "Optimal Dispatchable DG Location and Sizing with an Analytical Method, based on a New Voltage Stability Index," *International Journal of Research and Technology in Electrical Industry*. Journal homepage: ijrtei.sbu.ac.ir IJRTEI, vol. 1, no. 1, pp. 95–104, 2022, doi: 10.52547/ijrtei.1.1.95.
7. M. I. Akbar, S. A. A. Kazmi, O. Alrumayh, Z. A. Khan, A. Altamimi, and M. M. Malik, "A Novel Hybrid Optimization-Based Algorithm for the Single and Multi-Objective Achievement with Optimal DG Allocations in Distribution Networks," *IEEE Access*, vol. 10, pp. 25669–25687, 2022, doi: 10.1109/ACCESS.2022.3155484.
8. A. R. Gupta and A. Kumar, "Deployment of Distributed Generation with D-FACTS in Distribution System: A Comprehensive Analytical Review," *IETE Journal of Research*, pp. 1–18, Jul. 2019, doi: 10.1080/03772063.2019.1644206.
9. C. M. Castiblanco-Pérez, D. E. Toro-Rodríguez, O. D. Montoya, and D. A. Giral-Ramírez, "Optimal placement and sizing of d-statcom in radial and meshed distribution networks using a discrete-continuous version of the genetic algorithm," *Electronics (Switzerland)*, vol. 10, no. 12, Jun. 2021, doi: 10.3390/electronics10121452.
10. Sabarinath. G and T. G. Manohar, "Application of Bird Swarm Algorithm for Allocation of Distributed Generation in an Indian Practical Distribution Network," *International Journal of Intelligent Systems and Applications*, vol. 11, no. 7, pp. 54–61, Jul. 2019, doi: 10.5815/ijisa.2019.07.06.

11. B. Pottukkannan, Y. Thangaraj, D. Kaliyaperumal, M. I. Abdul Rasheed, and M. Petha Perumal, "Integration of solar and wind based dgs with dstatcom in distribution systems using modified bat algorithm," *Gazi University Journal of Science*, vol. 32, no. 3, pp. 895–912, 2019, doi: 10.35378/gujs.358228.
12. A. A. Ghavifekr, A. Mohammadzadeh, and J. F. Ardashir, "Optimal Placement and Sizing of Energy-related Devices in Microgrids Using Grasshopper Optimization Algorithm," Feb. 2021. doi: 10.1109/PEDSTC52094.2021.9405951.
13. P. Prakash, "Optimal DG Allocation Using Particle Swarm Optimization," in *Proceedings - International Conference on Artificial Intelligence and Smart Systems, ICAIS 2021*, Mar. 2021, pp. 940–944. doi: 10.1109/ICAIS50930.2021.9395798.
14. A. R. Gupta and A. Kumar, "Annual energy savings with multiple DG and D-STATCOM allocation using PSO in DNO operated distribution network," in *Advances in Intelligent Systems and Computing*, vol. 698, Springer Verlag, 2019, pp. 1–10. doi: 10.1007/978-981-13-1819-1_1.
15. R. Vempalle and P. K. Dhal, "Optimal placement of distributed generators for maximum savings using PSO-SSA optimization algorithm," in *Proceedings of the Confluence 2021: 11th International Conference on Cloud Computing, Data Science and Engineering*, Jan. 2021, pp. 624–629. doi: 10.1109/Confluence51648.2021.9377189.
16. S. Salkuti, "Optimal Allocation of DG and D-STATCOM in a Distribution System using Evolutionary based Bat Algorithm," *International Journal of Advanced Computer Science and Applications*, vol. 12, Dec. 2021, doi: 10.14569/IJACSA.2021.0120445.
17. D. B. Prakash and C. Lakshminarayana, "Multiple DG placements in radial distribution system for multi objectives using Whale Optimization Algorithm," *Alexandria Engineering Journal*, vol. 57, no. 4, pp. 2797–2806, Dec. 2018, doi: 10.1016/j.aej.2017.11.003.
18. A. Eid, S. Kamel, H. M. Zawbaa, and M. Dardeer, "Improvement of active distribution systems with high penetration capacities of shunt reactive compensators and distributed generators using Bald Eagle Search," *Ain Shams Engineering Journal*, vol. 13, no. 6, Nov. 2022, doi: 10.1016/j.asej.2022.101792.
19. K. S. Sambaiah, "Renewable energy source allocation in electrical distribution system using water cycle algorithm," *Materials Today: Proceedings*, vol. 58, pp. 20–26, 2022, doi: 10.1016/j.matpr.2021.12.569.
20. A. Eid, "Allocation of distributed generations in radial distribution systems using adaptive PSO and modified GSA multi-objective optimizations," *Alexandria Engineering Journal*, vol. 59, no. 6, pp. 4771–4786, Dec. 2020, doi: 10.1016/j.aej.2020.08.042.
21. R. Kandpal and A. Sharma, "Impact of Lockdown on Operation of Distribution System with Renewable Energy Sources and D-STATCOM," in *2022 IEEE Delhi Section Conference (DELCON)*, Feb. 2022, pp. 1–7. doi: 10.1109/DELCON54057.2022.9752949.

22. A. Ranjan Srivastava, M. Khan, F. Y. Khan, and S. Bajpai, "Role of Renewable Energy in Indian Economy," in *IOP Conference Series: Materials Science and Engineering*, Oct. 2018, vol. 404, no. 1. doi: 10.1088/1757-899X/404/1/012046.
23. S. Singh, M. Singh, and S. C. Kaushik, "Feasibility study of an islanded microgrid in rural area consisting of PV, wind, biomass and battery energy storage system," *Energy Conversion and Management*, vol. 128, pp. 178–190, Nov. 2016, doi: 10.1016/j.enconman.2016.09.046.
24. "National Renewable Energy Laboratory (USA). https://www.nrel.gov/research/data-tools.html".
25. K. Mahesh, P. Nallagownden, and I. Elamvazuthi, "Optimal placement and sizing of distributed generators for voltage-dependent load model in radial distribution system," *Renewable Energy Focus*, vol. 19–20, pp. 23–37, Jun. 2017, doi: 10.1016/j.ref.2017.05.003.
26. V. V. S. N. Murty and A. Kumar, "Mesh distribution system analysis in presence of distributed generation with time varying load model," *International Journal of Electrical Power and Energy Systems*, vol. 62, pp. 836–854, 2014, doi: 10.1016/j.ijepes.2014.05.034.
27. Y. M. Atwa, E. F. El-Saadany, M. M. A. Salama, and R. Seethapathy, "Optimal renewable resources mix for distribution system energy loss minimization," *IEEE Transactions on Power Systems*, vol. 25, no. 1, pp. 360–370, Feb. 2010, doi: 10.1109/TPWRS.2009.2030276.
28. Albadi, Mohammed, El-Saadany, and Ehab, "Demand Response in Electricity Market: An Overview."
29. S. Widergren, C. Marinovici, T. Berliner, and A. Graves, "Real-time pricing demand response in operations," 2012. doi: 10.1109/PESGM.2012.6345195.
30. A. K. Lal Karn and S. Kakran, "Operation Management of Microgrid Supplying to the Residential, Industrial and Commercial Community using Different Demand Response Techniques," Jun. 2022, pp. 1–6. doi: 10.1109/icepe55035.2022.9798064.
31. N. G. Paterakis, O. Erdinç, and J. P. S. Catalão, "An overview of Demand Response: Key-elements and international experience," *Renewable and Sustainable Energy Reviews*, vol. 69. Elsevier Ltd, pp. 871–891, Mar. 01, 2017. doi: 10.1016/j.rser.2016.11.167.
32. "Live hourly pricing." https://hourlypricing.comed.com/live-prices/
33. A. Khurana, V. V. R. Kumar, and M. Sidhpuria, "A Study on the Adoption of Electric Vehicles in India: The Mediating Role of Attitude," *Vision*, vol. 24, no. 1, pp. 23–34, Mar. 2020, doi: 10.1177/0972262919875548.
34. L. Sørensen, K. B. Lindberg, I. Sartori, and I. Andresen, "Analysis of residential EV energy flexibility potential based on real-world charging reports and smart meter data," *Energy and Buildings*, vol. 241, Jun. 2021, doi: 10.1016/j.enbuild.2021.110923.
35. Sandeep Dhundhara, Y.P Verma "Application of Micro Pump Hydro Energy Storage for Reliable Operation of Microgrid System", *IET Renewable Power Generation*, vol. 14, Issue 8, 2020.

36. S. Dhundhara, Y. P. Verma, and A. Williams, "Techno-economic analysis of the lithium-ion and lead-acid battery in microgrid systems," Energy Convers. Manag., vol. 177, pp. 122–142, 2018
37. B. Singh and A. K. Sharma, "Network Constraints Economic Dispatch of Renewable Energy Sources with Impact of Energy Storage," *International Journal of Computing and Digital Systems*, vol. 11, no. 1, pp. 423–440, 2022, doi: 10.12785/ijcds/110135.
38. C. Sabillon, A. A. Mohamed, B. Venkatesh, and A. Golriz, "Locational Marginal Pricing for Distribution Networks: Review and Applications."
39. S. Mirjalili, S. M. Mirjalili, and A. Lewis, "Grey Wolf Optimizer," *Advances in Engineering Software*, vol. 69, pp. 46–61, 2014, doi: 10.1016/j.advengsoft.2013.12.007.
40. J. H. Teng, "A direct approach for distribution system load flow solutions," *IEEE Transactions on Power Delivery*, vol. 18, no. 3, pp. 882–887, Jul. 2003, doi: 10.1109/TPWRD.2003.813818.

11

Energy Storage Systems and Charging Stations Mechanism for Electric Vehicles

Saurabh Ratra[1]*, Kanwardeep Singh[2] and Derminder Singh[1]

[1]Department of Electrical Engineering and Information Technology, Punjab Agricultural University, Ludhiana, Punjab, India
[2]Department of Electrical Engineering, Guru Nanak Dev Engineering College, Ludhiana, Punjab, India

Abstract

This chapter focuses on energy storage by electric vehicles and its impact in terms of the energy storage system (ESS) on the power system. Due to ecological disaster, electric vehicles (EV) are a paramount substitute for internal combustion engine (ICE) vehicles. However, energy storage systems provide hurdles for EV systems in terms of their safety, size, cost, and general management issues. Furthermore, focusing solely on EVs is insufficient because electrical vehicle charging stations (EVCS) are also required for the deployment of these vehicles. Because these vehicles are powered by electricity, installing these charging stations presents some challenges. Grid overloading and load forecasting were previously major issues. The latter refers to charging time and charging station traffic management. This chapter discusses the essential terms of charging stations (CS). To address these issues, various technologies are discussed, including a brief overview of lithium-ion battery charging techniques and battery management system (BMS). As the Indian government is focusing on creating an an eco-friendly system, and as part of its operation is to condense CO_2 emissions from the transportation segment, the organization of EVs and the installation of electric vehicle charging stations (EVCS) is of utmost importance. The government has already minimized taxation on EVs and also provides subsidies for CS installation. As a result, in this context, different procedures issued by the government of India are deliberated which assist an individual in installing CS at their location.

---

*Corresponding author: saurabhratra@pau.edu

Sandeep Dhundhara, Yajvender Pal Verma, and Ashwani Kumar (eds.) *Energy Storage Technologies in Grid Modernization*, (317–340) © 2023 Scrivener Publishing LLC

Keywords: Storage technologies, charging stations, charging structures, electric vehicle, power system, smart grid, transport

11.1 Introduction to Electric Vehicles

With the increase in CO_2 emissions and oil scarcities around the globe, a dramatic transition in the conveyence sector is reported, with an internal combustion engine (ICE) based conveyence giving way to electric vehicles (EVs). EVs are the ideal alternative because oil is the major source for ICE vehicles, which is the fundamental cause of worldwide ecological disaster [1]. The EV uses plug-in recharged storage to function on electricity from batteries, fuel cells, and ultra-capacitors; the final supply of electricity comes from power plants and renewable energy sources. EVs use thermoelectric generators and regenerative braking to cut down on energy waste. Whereas a thermo-electric generator automatically converts heat from engine to electricity, the braking mechanism of the vehicle captures this energy, and converts it back to electrical energy, thereby returning to batteries. Thus EVs rely heavily on the energy storage technologies that are currently available.

11.1.1 Role of Electric Vehicles in Modern Power System

Substituting ICE with electrical motors, these vehicles may deliver additional benefits such as fewer moving parts, extraordinary torque, huge power density, and improved efficiency, among others [2]. Electrical batteries are a major source of power for motors. The main problem for EVs is battery charging in a short amount of time, which means that in turn EV charging stations become handy. Depending on the level of charging, charging stations can be of several sorts [1]. Various standards codes, i.e., the Society for Automobile Engineers (SAE), etc., and the International Organization for Standardization, provide strategies for charging vehicles based on their rated capacity [1]. The deployment of EVCS comes with several problems. The number of vehicles on the road is growing every day, necessitating high electrical power which puts pressure on the system to generate additional electrical power that may overload the system and force power generation to expand, which, if done using fossil fuels, would be just as harmful to the atmosphere [3]. Moreover, grid congestion may result in voltage regulation issues, a loss in system dependability and efficiency, an increase in thermal loading, and the most significant impact on load forecasting. In the electrical distribution system, load forecasting is critical for predicting power generation by evaluating peak and base loads [1, 3]. However, introduction of EVs and electric vehicle charging stations (EVCS), load forecasting has become more complex, as the calculation of variable loads has proven to be the most difficult task. Moreover, the (EV) relies on plug-in rechargeable storage to run on battery power and is highly dependent on energy storage technologies.

11.1.2 Various Storage Technologies

Electric energy storage needs for EVs are taken into account in a number of ways. The key components for effective energy storage are supervision system, power electronics interfaces, safety-related power conversion as well as protection. Figure 11.1 shows EV architecture [3]. Figure 11.2 depicts the HEV series-parallel configuration. The assortment and supervision of energy storage, and supervision of storage system are important for EV future technologies. Managing energy resources, selecting ESSs, and avoiding anomalies are necessary for providing sophisticated features in an EV. In this chapter, the main aim is to include the current state of EVCS, and ESSs, their updated features, challenges, and difficulties with current systems.

Renewable energy has helped off-grid power users with ESSs during the last few decades. In that regard, EVs are developing technologies that use ESS to replace fossil fuels with energy resources obtained from renewable energy sources [4].

The utilization of energy in a certain form is used to categories ESS systems. Different categories such as electrical and chemical for storage

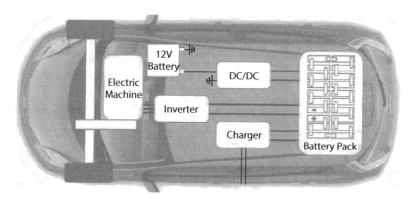

Figure 11.1 Battery-powered electric vehicle architecture.

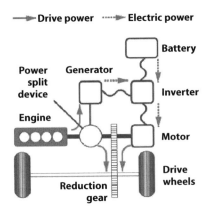

Figure 11.2 Series parallel configuration of HEV.

systems are available. These systems are divided into a number of categories based on how they were formed and what materials they were composed of [5, 6].

Around the world, mechanical-based storage systems are frequently utilized to generate power. The three mechanical storage systems are packed hydro storage (PHS), compressed air energy storage (CAES), and flywheel energy storage (FES). This storage technology accounts for approximately 99% of the world's electric storage capacity or approximately 3% of global power generation capacity [7].

EVs and power systems can utilise flywheel energy storage (FES) devices thanks to advancements in power electronics. The effectiveness and evaluated power of FESs, respectively, range from 90% to 95% and 0 to 50 MW [8]. Through the use of a transmission device, the energy maintained by the continuously rotating flywheel is transformed into electrical energy. Electro-chemical-based energy storage includes all typical rechargeable batteries (EcSSs). EcSSs use a reversible mechanism with little physical changes and high energy efficiency is required to alter chemical energy to electrical energy and vice versa [9]. However, reversible mechanism may reduce cell life.

Flow Batteries (FBs) may be recharged, and they include electroactive substances that store energy. Chemical energy is converted into electric energy by dissolving electroactive species in the liquid electrolyte in tanks and pumping the liquid through an electrochemical cell. Redox flow (RFB) and hybrid flow (HFB) are two examples of FBs [6]. The entire volume of RFB tank determines battery's overall energy [6]. Portable energy storage for EVs is dominated by secondary batteries (SBs). These batteries generate electricity through an electrochemical reaction mechanism and stock electrical power in the form of chemical energy [6]. Good qualities of SB include its wide temperature performance range, high specific energy, and other components. Toxic elements are present in most batteries. Consequently, the environmental impact of battery disposal must be taken into account.

Chemical storage systems (CSSs) contain chemical substances that react chemically to produce other molecules while storing and releasing energy [10]. The FC is a common chemical storage device that continually transforms fuel's chemical energy into electrical energy. The manner in which

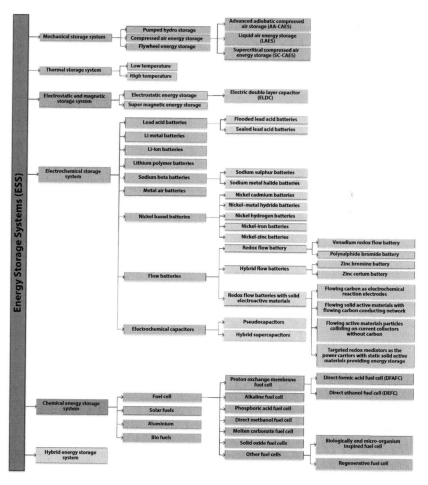

Figure 11.3 Organization of different energy storage technologies.

they provide energy sources is the primary distinction between an FC and a battery. To create energy in FC, energy and required oxidants are provided from outside, and these components are built within the battery. Figure 11.3 represents the ESS organization in detail [3].

11.1.3 Electric Vehicle Charging Structure

In practically every market, plug-in electric vehicles (PEV) are still in the initial phases of progress. The absence of a public charging infrastructure continues to be a deterrent to PEV adoption. Key statistics, particularly the ratio of PEVs to public charging sites, are frequently used to estimate the future demands for charging infrastructure. In this chapter, the review has been reported regarding the foundation for the medium- to long-term requirement for infrastructure for charging.

Sundararajan *et al.* [11] have created a novel charging architecture for EVs, comprised of sensors that can interface with vehicle-to-grid (V2G) technology by establishing a communication link. This allows for the regular monitoring of load caused by car charging. V2G technology and the Intelligent Transportation System usage can also reduce waiting time at charging stations. Gupta *et al.* [12] addressed the IoT concept and automatic vehicle organization algorithm which uses the ICT technique to connect vehicles, grids, and charging stations on an internet protocol (IP) level (ICT). This may assist in reducing load congestion and long lines at EVCS. The major significant challenge in the expansion of EVs is the charging time. In these cars, lithium-ion batteries are utilised to store electrical energy, and these batteries employ graphite to store charge. Silicon is added to the graphite layer to boost the charge storage capacity of lithium-ion batteries. The batteries must be charged properly in order to have an enhanced performance. Because charging of lithium-ion battery is a delicate operation prone to overvoltage and overheating, there is a variety of charging strategies available, as reported in [13]. The different level charging scheme for lithium-ion batteries is discussed by Kodali *et al.* [14], in which the battery is charged with five dissimilar points of current. However, adequate observing of battery State of Charge (SOC) is essential while charging these batteries, which reveals how much of the battery is charged or depleted. Battery Management System (BMS) is utilized for process monitoring and smooth functioning [15–17]. The Internet of Things (IoT) can aid with load forecasting at the BMS and SOC levels, which can help with power quality and reliability issues.

11.2 Introduction to Electric Vehicle Charging Station

EVCS is a common point where EVs get charged provided with reliability, supervision, and also have conversion systems, for rapid charging through high voltage and high current. The following are certain basic EVCS terms.

11.2.1 Types of Charging Station

(i) Residential Charging Station: Residential charging stations are critical for ushering in the EV age, as they will greatly lower the strain on the power system. Residential car charging can be accomplished by consuming minimum current from the system, assisting the system to meet demand required to serve additional electrical power required during peak load. According to [18], EV charging during the night is cheapest and has minimum effect on the system, which minimises the load on the system because during base load hours per unit cost of electricity is reduced for charging vehicles during the night. Level 1 charging in a residential charging station can charge vehicles in as little as 7 to 8 hours. The same is represented in Table 11.1.

(ii) Charging Station Parking: EV took time to charge, but putting that duration to good use while parking, you can reduce the strain on public charging stations and the power grid. Automobiles are parked at work for nearly five hours per day, according to the National Household Travel Survey [19]. The provision is accessible for commercial sites and other areas where sufficient charging infrastructure are available, in addition to workplaces [20, 21].

(iii) Public Charging Station: Public charging stations are intended to give vehicles quick charging. Traditional charging, on the other hand, takes additional time for battery charging. Different charging methodologies and fast charger configurations are used to achieve fast charging. A charging station's charger is typically composed of AC-DC converter at front side and DC-DC converter on the rear side. The DC link capacitor connects both converters [22]. MOSFET and IGBT are used in chargers with high-frequency switching

rectifiers, and they work based upon Pulse Width Modulation (PWM). The PWM technique delivers high-efficiency and precise power conversion [23]. For this, a high-quality filter is used to connect the charger to the power grid, preventing harmonics from affecting the grid and the motor attached to the car [22, 24].

(iv) *Battery Swapping Station*: The Battery Swapping Station (BSS) was created to highlight the charging time problem and the necessity for a charged car. The exhausted battery or battery pack in a BSS car can be swapped out for a fully charged one right away, saving the time spent waiting for the vehicle's battery to charge. The BSS looks after the battery life through BMS monitor [25]. There are certain challenges to sort when deploying a BSS. An important challenge which was addressed is the battery pack's design which is easy to remove and reattach to cars. Brand compatibility is a different issue of the battery packs. Manufacturers can generate substitutable battery packs for BSS and EVs by using a common standard structure. There is also the issue of battery degradation and ownership, which is the most significant impediment to BSS technology.

11.2.2 Charging Levels: The vehicle charging rate is presented in Table 11.1 [1] based on didfferent international standards and procedures for charging of vehicles.

11.2.3 EV Charging: Based on energy transfer mode, EV charging systems are divided into two types: conductive and inductive charging systems.

(i) Conductive Charging: In this process, the system utilises a cable or connector for direct connection between vehicle and charger. It is a basic charging infrastructure at present. Based on type of charging, it is either on-board or off-board charging. This particular charging is highly efficient; moreover, every EV manufacturer provides this service. Vehicles with this charging approach are currently available, which includes Chevrolet Volt and Mitsubishi i-MiEV [1, 13].

(ii) Inductive Charging: This charging provides wireless charging, which is a new emerging concept. No physical contact is

Table 11.1 Description of different EV chargers and charging levels.

EV charging levels	Connection segment	Charger type	Utilization and location	Power consuption (kW) and current capacity (A)
Society for Automobile Engineers AC and DC charging standards				
Charging Level 1 AC 230V (EU)/120V(US)	1-Φ	Boarding	Residence cum Office	1.6kW/12.5A, 1.95kW/20.25A
Charging Level 2 400V (EU) = 240V (US)	1-Φ/3-Φ	Boarding	Public cum Private	4.2kW/17.4A, 8.2kW/32.1A
Charging Level 3 210-599 V	3-Φ	Off-Boarding	Commercial cum Filling Station	Station 50.01kW, 100.02kW
DC Charging Level 1/200-450V		Off-Boarding	Committed Charging Stations	40.02kW/80.4A
DC Charging Level 2= 200-400V		Off-Boarding	Committed Charging Stations	90.01kW/200.05A
DC Charging Level 3= 200-600V		Off-Boarding	Committed Charging Stations	240.04kW/400.4A

(Continued)

Table 11.1 Description of different EV chargers and charging levels. (*Continued*)

EV charging levels	Connection segment	Charger type	Utilization and location	Power consuption (kW) and current capacity (A)
AC and DC Charging standards for IEC				
Charging Level 1 AC	1-Φ	Boarding	Residence cum Office	4.5-7.55kW/16.5A
Charging Level 2 AC	1-Φ/3-Φ	Boarding	Public cum Private	8.2-15.1kW/32.4A
Charging Level 3 AC	3-Φ	Boarding	Commercial cum Filling Station	60.04-120.01kW/250.04A
DC Fast Charger		Off-Boarding	Committed Charging Stations	1000.04-2000.05kW/400.04A
Charging Standard CHAdeMo				
DC Fast Charger		Off-Boarding	Committed Charging Stations	62.57kW/125.4A

required between vehicle and charger. Electromagnetic induction principle is applicable as that to transformers [13, 26]. To transfer energy through thin air, a magnetic field is used. The only disadvantage with this approach is that it has lower efficiency and power density in comparison to the above-mentioned charging and is also cost inefficient [1]. Vehicles can be charged while running if charging strips are placed along the highway. The term for this type of charging is dynamic wireless charging. Roads that can supply electrical supply to vehicles via wireless power technique (WPT) can be referred to as electrified roads [1, 27]. Charging of vehicle when driving reduces the vehicle charging time [1]. The method utilized in different countries, along with more research, are being worked upon to enhance the efficiency [28].

11.2.4 Charging Period: Charging period is another major task for EV technology. The battery recharging time in EV is greater as compared to oil refuelling time. Primarily there are five major factors which affect system fast charging to minimise charging time [29].

- Sizing of battery: Charging time increases as the volume measured in kWh increases. A large amount of time is required to charge the battery.
- State of charge (SOC): Battery state of charge (SOC) determines whether it is fully charged, fully discharged, or partially discharged, and therefore charging period differs accordingly.
- Vehicle charging rate: The vehicle can only be charged at the extreme amount and not any higher. A battery with a maximum rate of charge of 30kW cannot be charged by means of a 60kW charger.
- Charge point pricing: The rating of the outlet to which the battery is connected determines the charging time. If you charge a 30kW battery with a 10kW outlet, it will charge at the same rate as a 10kW battery, resulting in a longer charging time [29].

Table 11.2 Vehicle time of chargers.

Battery specification	Capacity (kWh)	Range (miles)	Charging period (hours)				
			3.7 kW	7 kW	22 kW	43-50 kW	150 kW
Model I	40.5	145	12	7	7	2	NA
Model II	75.08	240	22	12	6	3	2
Model III	14	25	5	5	5	1	NA

When charged with a 10kW charger, the charging time is less than 8 hours, and the battery can travel 50-60 miles. Table 11.2 demonstrates the amount of time mandatory to charge the batteries of various corporation. Model I is the 2018 Nisan Leaf, Model II is the 2019 Tesla Model S, and Model III is the 2018 Mistubishi Outlander PHEV.

11.3 Modern System Efficient Approches

The placement of a huge number of CS increases electrical power demand. Since large power is taken from the generating source, the grid may get overloaded, causing different power-related issues such as voltage fluctuation, voltage regulation issues, etc. These problems have an impact on the inclusive efficiency of the system that is unacceptable for EV and EVCS development. To address these issues, different approaches are presented as below.

11.3.1 Smart Grid Technology

The use of smart grid technology can help to alleviate the problem of uncoordinated power supply and reliability. The smart grid established a communication channel between the grid and the user to ensure proper load monitoring based on the area [30]. Because remote terminal units are installed at each feeder, they send information about any fault conditions as well as power usage at each feeder; smart grid implementation ensures the system's safe operation. This technique can provide the grid with load information in advance, ensuring smooth generation and, as a result, no reliability issues [30, 31]. Furthermore, this technique is useful in load forecasting and is linked to the vehicle so that the SOC of a battery can be shared with the grid and the grid can trace the nearby charging station from which feeder will take the load [34].

11.3.2 Renewable Energy Rechnology

Fossil fuels are both the primary cause of environmental degradation and the primary source of electricity generation today. More electrical power is required as electric vehicle adoption grows. As a result, using these fossil fuels in a different way to meet needs is not a wise decision. Renewable energy sources are the best option for charging electric vehicles because they reduce both carbon emissions and grid load [32, 33]. Solar energy is the most basic and least expensive renewable energy source available in many parts of the world today. Installing a solar power plant at home is also the most straightforward and secure method of obtaining electricity. To reduce direct load on the grid, solar panels can be installed on the roofs of public EVCS, shopping malls, offices, and other large-surface-area buildings [33]. Although the initial cost of installing a solar plant on a roof is high, the operating costs are very low, lowering the operating costs of EVs in comparison to ICE-powered vehicles.

11.3.3 V2G Technology

To maintain power system balance, active power and frequency must be balanced, as overloading and underloading can cause a frequency mismatch, affecting system stability. As a result, a bidirectional energy flow system is recommended, in which the grid supplies power to the vehicle and the vehicle feeds power back to the grid when not in use. This system is also referred to as a vehicle-to-grid (V2G) system [34]. According to statistics, 90% of electric vehicles are idle every day, and they can contribute to meeting the high energy demand by supplying power back to the grid [35]. To maintain a balanced load, automatic generation control (AGC) was introduced to control the modern power network. Traditional AGC regulates the generating unit in response to load changes. Residential small solar plants that are not currently charging vehicles can help to reduce grid load by generating power that can be fed back into the grid. This will be a clean, green, and cost-effective form of electricity [18, 35].

11.3.4 Smart Transport System

To enhance the system's acumen, an intelligent transportation system (ITS) has been introduced. It is made up of sensors, actuators, and an embedded processor that aids in tracking the specific area's traffic congestion. In other words, it establishes an open communication channel between two or more people using the parking lot or charging station. Using the internet as a medium or cellular network with on-board geographic information Geographic Information Systems (GIS), Global Positioning Systems (GPS), and advanced traffic flow modelling techniques this correspondence can be easily accomplished [12, 36]. The process can be monitored and controlled using the Internet of Things (IoT). IoT is also useful in determining the SOC of the EV battery and transmitting this information to the grid so that proper load monitoring can be carried out. Apart from that, using this technology it is easy to pre-book a slot at a charging station and check the status of an empty slot [12, 18, 36]. This technique improved load forecasting and allowed for easy communication with renewable energy-based generating plants located at home, office, parking lot, shopping mall, or charging station [12].

11.4 Battery Charging Techniques

While considering energy density and EVs permanence, lithium-ion batteries are presently the most well-known batteries. These are built through different cells which are associated through series connection and then through parallel to make a segment, and different segments are associated in series to make a single source of battery. The use of different cells aids in battery upkeep and observing [37]. The most difficult challenge is the efficient and quick charging of these batteries. Different charging techniques of the battery are mentioned below.

(a) Continuous current charging
Battery charging is done through constant current in this scheme. If a higher quantity of cells are connected, this approach will not be successful as it can affect cell balancing problem. As a result, this approach is quite incompetent and may cause stress in the cells [14]. To use this approach, a large amount of current is obligatory, which will certainly do the fast charging of the battery, but the temperature of the battery will increase, potentially leading to the battery's sudden death [14, 15].

(b) Constant voltage constant current charging
There are four charging modes in this scheme [14].

- Pre-Charge Mode: To avoid cell overheating, nearly 9% of the battery is charged at maximum current in this mode.

- Constant Current Mode: Until it reaches 4.2 V, the battery operates at less than 1 C per second. This voltage will rise.
- Constant Voltage Mode: Battery charging at 4.2V constant voltage till it is fully charged. Because the constant current method causes overheating, the battery will be charged by constant voltage even though it is fully charged.
- Charge Termination Mode: The charge is terminated by monitoring the charge current and terminating it when it reaches 0.02-0.07C using the minimum charge current method.

Despite these benefits, this method is still not widely used as it requires a higher period for charging the battery completely.

(c) Scheme for charging in stages
All of the methods mentioned above have the disadvantage of taking longer to charge. However, there is also battery charging for different current profiles, each within voltage limits. This approach was developed because large current can raise the battery temperature [14]. When the internal resistance of the battery is low, a high current is required to charge the battery. Charging currents and threshold voltage limits are determined by the charging rate and SOC. This method charges batteries faster and more efficiently while preserving battery life [16].

11.4.1 Electric Vehicle Charging Station in Modern Power System

EV charging can be accomplished using AC or DC power supplies. Based on the country's electrical system, AC charging has varying voltage and frequency levels. AC charging is classified into three voltage level 1, level 2, and level 3. Level 1 and level 2 CS may be placed in a secluded location, whereas Level 3 CS require distinct electrical setup, require consent from power providers and are typically manufactured for public CS. DC charging is faster and has larger capacity to charge at the same voltage level. EV can be charged minmum to minimum in 20 minutes. However, different charging modes are available for EV charging in modern power systems.

11.5 Indian Scenario

The government is putting additional emphasis on reducing CO_2 emissions and is doing every required phase. India is in the top ten nations in terms of automobile arcade size [38]. As a result, the use of more and more oil-based vehicles has conflicting influence on the atmosphere. The National Electric Mobility Mission Plan (NEMMP) 2020 [39] estimates that 7 to 10 million electric vehicles will be on the road by 2025, reducing vehicle carbon emissions by 1.3%. According to the Indian government's vision for renewable energy generation, only clean energy sources can generate around 175GW of power by 2022, with solar accounting for 100GW of that total [40, 41].

Major guidelines for installing a public charging station are provided by the Ministries of Power and Housing and Urban Affairs [42–44] as follows:

- It is a prohibited activity, which means that anyone can set up a charging station. However, the individual must notify the electricity distributor in order to obtain a proper electricity supply.
- Charging station may borrow electrical power from any generation company, either by their own power generation or through the use of solar panels, windmills, or other means [45].
- The station for charging requires a special transformer with all protection devices coupled to the substation.
- There is enough space inside the station for vehicles to manoeuvre.
- Civil and firefighting work should be done properly.
- For the installation of a public charging station, a minimum set of 5 charging points as shown in Table 11.3 are required. There is no requirement to use only the chargers listed in Table 11.3. Owners are free to use any connector they want, as long as it meets the same standards and specifications as these chargers and the BIS standards. The number of charging points can be increased based on the number of EVs, but a minimum of 5 chargers is required.
- In terms of distance, it is recommended that charging stations be placed every 3 kilometers in cities and every 25 kilometers on highways. This distance can be reduced by adding more stations but not increased [46].

Table 11.3 Different types of EV chargers.

Charger type	Charger connection	Minimum electrical power (kW)	Voltage (V)	No. of charging points
Fast	CCS	60	230-1100	2
	CHAdeMO	60	230-1100	2
	AC Type 2 AC	30	350-450	2
Slow/ Moderate	001 DC	20	70-210	2
	001 AC	15	220	4

11.6 Energy Storage System Evaluation for EV Applications

Specific characteristics are used to evaluate ESSs for EV applications as reported above as well as the required demand for EV charging. Figure 11.4 demonstrates the operating time of various ESSs based on their power releases. Figure 11.5 depicts the applications of various ESSs as the demand for EVs and other modes of transportation generally requires. For EV

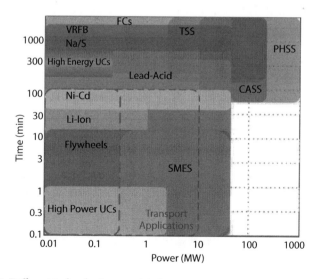

Figure 11.4 Different technologies associated to energy storage.

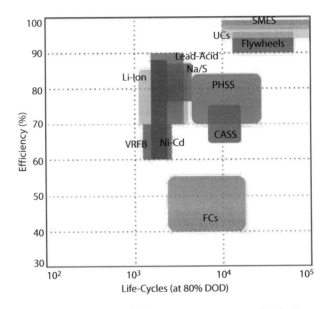

Figure 11.5 Efficiency distribution of different energy storage technologies.

applications, a power range of 10 kW to several hundred kW is required for a few hours of operation [47, 48].

As shown in Figure 11.5, the ESS technologies distribution can be assessed through efficiency and predictable life cycle. SMESs and flywheels has relatively large efficiencies and a enhanced life cycle at 80% DOD. Li-ion, Ni-Cd batteries have efficiencies around 70-85% and a life cycle of 2000-4500 cycles. ESS techniques are crossbred by merging batteries with UCs, flywheels, as well as the growth and obtainability of standard EVs for next-generation mode of conveyance [49].

11.7 ESS Concerns and Experiments in EV Solicitations

The current state of ESS development is adequate for EV energy storage and powering. These applications, however, continue to face issues such as raw material support and disposal, energy management, power electronics interface, sizing, safety measures, and cost. The following sections address the key issues and make recommendations.

11.7.1 Raw Materials

The availability of raw materials and supplies for the production of ESSs and the development of related products is a difficult issue. Electrode, electrolyte, separator materials, and chemical solutions for batteries, UCs, and FCs; flywheel materials; superconducting materials for SMES; and hydrogen fuel for FCs are critical components in ESS manufacturing. The most significant advancement in ESS design and technology for EV applications is the consideration of high-grade ESS materials, alloys, and solution preparation, as well as the use of ESSs with high charge capacities. Current and future research considers recycling, refurbishing, and reusing used ESS materials [50].

11.7.2 Interfacing by Power Electronics

Unspecified and unorganized power storage and distribution may reduce ESS performance, life cycle duration, and efficiency, as well as cause extreme power loss and abuse, unexpected explosions and damages, and restricted load behavior and life. The power electronics interface deals with situations involving power conditions, controls, and conversions for storing and supplying ESS and load requirements in order to optimize the system's overall performance, durability, and efficiency. For power conversion, power flow control, power management control, motor drive, energy management, charge balancing, and safe operation, ESSs in EV applications require a power electronics interface [51].

11.7.3 Energy Management

After each life cycle, EESs must be recharged using either ultimate or temporary energy resources. An energy management system (EMS) manages all possible energy resources for powering EV ESSs. Energy resource systems, ESSs, and power electronics are all dealt with by EMS. Grid power, solar energy, hydrogen energy, regenerative braking, thermal energy, vibration energy, flywheel system, SMES, and other energy sources are all possibilities for recharging ESSs in EVs. Modern EV systems are designed to effectively and intelligently manage all energy resources [52].

11.7.4 Environmental Impact

Despite EV usage, demand for oil has been significantly reduced, EV ESSs have had little influence on environmental pollution during manufacturing, disposal, and recycling of ESSs. Furthermore, the dispensation and manufacturing of ESS causes respiratory, pulmonary, and neurological problems. As a result, safety measures and sophisticated tools are critical in managing the entire production and maintenance processes of ESSs, particularly in EV applications [53].

11.7.5 Safety

Safety measures ensure that ESSs operate at demand rates while also improving their lives and performances. Li-ion batteries require protection from overcharging and over-discharging in EV applications, Zn-Air batteries require protection from short-circuit, Na-S batteries require safety from high-temperature and ZEBRA batteries require thermal management system. Power electronics interfaces are used in modern EVs for power management, power convertors, and controller to institute effective facilities and benign ESS operations [54].

11.8 Conclusion

Considering that traditional ICEs vehicles donate truncated efficiency and increased CO_2 and greenhouse gases emissions, the EV approach offers substitutes to ICE-based conveyence.

However, it is impossible to create EV systems without taking energy storage technology into account. The ESS technologies and their designs are covered in this chapter along with a variety of features for EV storage systems. Furthermore, ESS technologies, efficiency, and characteristics of EVs are presented. Problems and difficulties with the ESS approach in EV solicitations are also covered. With advancements in technology, ESSs are maturing more and more. The correct disposal, power electronics interfacing, safety precautions, and pricing of ESS remain issues. For improving energy and power density, high-quality ESS constituents, and organic solutions could be optimized in ESS design for EV solicitations.

Moreover, organizing CS concerns are studied with respect to grid overloading and battery charging time, which is instigated by additional CS waiting at EVCS. Various charging methods are available, and the battery is the primary constituent of an EV which must be efficiently charged without any destruction. Multi-stage charging is highly desired for fast charging because it does not degrade the battery.

References

1. Aghajan-Eshkevari, Saleh, et al. "Charging and discharging of electric vehicles in power systems: An updated and detailed review of methods, control structures, objectives, and optimization methodologies." *Sustainability*, 2022, 14(4): 2137.
2. J. G. West, "DC induction, reluctance and pm motors for electric vehicles," *Power Engineering Journal*, 1994, 8(2):77–88.
3. S. F. Tie, C.W Tan, "A review of energy sources and energy management system in electric vehicles". *Renew Sustain Energy Rev* 2013;20:82–102.
4. D. Hardin, "Smart grid and dynamic power management. *Energy management systems*, Giridhar Kini (Ed.), InTech. Available from: http://www.intechopen.com/books/energymanagement-systems/smart-grid-and-dynamic-powermanagement; 2011. [30.6.2015].
5. KT Chau, YS Wong, CC Chan, "An overview of energy sources for electric vehicles. *Energy Convers Manag* 1999;40:1021–39.
6. Electrical Energy Storage. White paper. International Electrotechnical Commission. (IEC), Geneva, Switzerland; 2011.
7. A. G. Olabi, T. Wilberforce, M. A., Abdelkareem, & M. Ramadan, "Critical review of flywheel energy storage system. *Energies*, 2021, 14(8), 2159.
8. M. A. Hannan,. M. M. Hoque, A. Mohamed, & A. Ayob, "Review of energy storage systems for electric vehicle applications: Issues and challenges. *Renewable and Sustainable Energy Reviews*, 2017, 69, 771–789.
9. N. Hiroshima, H. Hatta, M. Koyama, J. Yoshimura, Y . Nagura, K. Goto, Y. Kogo, "Test of three-dimensional composite rotor for flywheel energy storage system. *Compos Struct* 2016;136:626–34.
10. I. Dincer, MA Rosen, *Thermal energy storage: systems and applications*, 2nd ed. USA: John Wiley & Sons, Ltd; 2011.
11. Sundararajan, Raghul Suraj, and M. Tariq Iqbal. "Dynamic Modelling of a Solar Energy System with Vehicle to Home and Vehicle to Grid Option for Newfoundland Conditions", *European Journal of Electrical Engineering and Computer Science*, 2021, 5(3): 45-53.
12. Gupta, Manik, et al. "Lightweight branched blockchain security framework for Internet of Vehicles", *Transactions on Emerging Telecommunications Technologies*, 2022: e4520.
13. F. Zhang, X. Zhang, M. Zhang, and A. S. Edmonds, "Literature review of electric vehicle technology and its applications," in *2016 5th International Conference on Computer Science and Network Technology (ICCSNT)*. IEEE, 2016, pp. 832–837.

14. S. P. Kodali and S. Das, "Implementation of five level charging scheme in lithium ion batteries for enabling fast charging in plug-in hybrid electric vehicles," in *2017 National Power Electronics Conference (NPEC)*. IEEE, 2017, pp. 147–152.
15. Y. Yin, Y. Hu, S.-Y. Choe, H. Cho, and W. T. Joe, "New fast charging method of lithium-ion batteries based on a reduced order electrochemical model considering side reaction," *Journal of Power Sources*, 2019, 423: 367–379.
16. Qin, Yudi, et al. "A rapid lithium-ion battery heating method based on bidirectional pulsed current: Heating effect and impact on battery life", *Applied Energy*, 2020, 280: 115957.
17. Wu, Sen-Tung, et al. "A fast charging balancing circuit for LiFePO4 battery", *Electronics*, 2019, 8(10): 1144.
18. Burkert, Amelie, and Benedikt Schmuelling, "Challenges of conceiving a charging infrastructure for electric vehicles-An overview", *2019 IEEE Vehicle Power and Propulsion Conference (VPPC)*. IEEE, 2019.
19. Visakh, Arjun, and Selvan Manickavasagam Parvathy, "Energy-cost minimization with dynamic smart charging of electric vehicles and the analysis of its impact on distribution-system operation", *Electrical Engineering*, 2022: 1–13.
20. J. Babic, A. Carvalho, W. Ketter, and V. Podobnik, "Evaluating policies for parking lots handling electric vehicles," *IEEE Access*, 2017, 6: 944–961.
21. Habib, Salman, et al. "Contemporary trends in power electronics converters for charging solutions of electric vehicles", *CSEE Journal of Power and Energy Systems*, 2020, 6(4): 911–929.
22. Yan, Xiangwu, et al. "Virtual synchronous motor based-control of a three-phase electric vehicle off-board charger for providing fast-charging service", *Applied Sciences*, 2018, 8(6): 856.
23. Liu, Guozhong, et al. "Charging station and power network planning for integrated electric vehicles (EVs)", *Energies*, 2019, 12(13): 2595.
24. M. Di Paolo, "Analysis of harmonic impact of electric vehicle charging on the electric power grid, based on smart grid regional demonstration project los angeles," in *2017 IEEE Green Energy and Smart Systems Conference (IGESSC)*. IEEE, 2017, pp. 1–5.
25. Soares, Filipe Joel, PM Rocha Almeida, and JA Pecas Lopes. "Quasi-real-time management of electric vehicles charging", *Electric Power Systems Research* 108 (2014): 293–303.
26. Habib, Salman, et al. "A Comprehensive Topological Assessment of Power Electronics Converters for Charging of Electric Vehicles", *Flexible Resources for Smart Cities*. Springer, Cham, 2021. 133–183.
27. Lin, Yuping, et al. "Multistage large-scale charging station planning for electric buses considering transportation network and power grid", *Transportation Research Part C: Emerging Technologies*, 2019, 107: 423–443.
28. N. Shinohara, "Wireless power transmission progress for electric vehicle in japan," in *2013 IEEE Radio and Wireless Symposium*. IEEE, 2013, pp. 109–111.

29. Liu, Yayuan, Yangying Zhu, and Yi Cui, "Challenges and opportunities towards fast-charging battery materials", *Nature Energy*, 2019, 4(7): 540–550.
30. Bilal, Mohd, and Mohammad Rizwan. "Integration of electric vehicle charging stations and capacitors in distribution systems with vehicle-to-grid facility." *Energy Sources, Part A: Recovery, Utilization, and Environmental Effects* (2021): 1–30.
31. Zhang, Wei, and Chunting Chris Mi. "Compensation topologies of high-power wireless power transfer systems", *IEEE Transactions on Vehicular Technology*, 2015, 65(6): 4768–4778.
32. Wu, Xiaohua, *et al.* "Stochastic control of smart home energy management with plug-in electric vehicle battery energy storage and photovoltaic array", *Journal of Power Sources*, 2016, 333: 203–212.
33. Domínguez-Navarro, J. A., *et al.* "Design of an electric vehicle fast-charging station with integration of renewable energy and storage systems.", *International Journal of Electrical Power & Energy Systems*, 2019, 105: 46–58.
34. Peng, Chao, Jianxiao Zou, and Lian Lian. "Dispatching strategies of electric vehicles participating in frequency regulation on power grid: A review", *Renewable and Sustainable Energy Reviews*, 2017, 68: 147–152.
35. Bae, Youngsang, Trung-Kien Vu, and Rae-Young Kim. "Implemental control strategy for grid stabilization of grid-connected PV system based on German grid code in symmetrical low-to-medium voltage network", *IEEE Transactions on Energy Conversion*, 2013, 28(3): 619–631.
36. Ratra, S., Singh, D., Bansal, R. C., & Naidoo, R. M. (2021, December). Stochastic Estimation and Enhancement of Voltage Stability Margin considering Load and Wind Power Intermittencies. In *2021 IEEE 6th International Conference on Computing, Communication and Automation (ICCCA)* (pp. 744–748). IEEE.
37. Wu, H., Pang, G. K. H., Choy, K. L., & Lam, H. Y., "An optimization model for electric vehicle battery charging at a battery swapping statio", *IEEE Transactions on Vehicular Technology*, 2017. 67(2): 881–895.
38. Akhtar, Nadeem, and Vijay Patil. "Electric Vehicle Technology: Trends and Challenges", *Smart Technologies for Energy, Environment and Sustainable Development*, 2022, 2: 621–637.
39. Azadfar, Elham, Victor Sreeram, and David Harries. "The investigation of the major factors influencing plug-in electric vehicle driving patterns and charging behaviour", *Renewable and Sustainable Energy Reviews*, 2015, 42: 1065–1076.
40. Bandyopadhyay, Santanu, "Renewable targets for India", *Clean Technologies and Environmental Policy*, 2017, 19(2): 293–294.
41. Rubio-Aliaga, Álvaro, *et al.* "Multidimensional analysis of groundwater pumping for irrigation purposes: Economic, energy and environmental characterization for PV power plant integration", *Renewable Energy*, 2019, 138: 174–186.

42. M. Patrick, P. Weldon, and M. O'Mahony, "Future standard and fast charging infrastructure planning: An analysis of electric vehicle charging behaviour", *Energy Policy,* 2016, 89: 257–270.
43. Alosaimi, Wael, *et al.* "Toward a Unified Model Approach for Evaluating Different Electric Vehicles." *Energies,* 2021, 14(19): 6120.
44. "Guidelines for Implementation of Scheme for Farmers for Installation of Solar Pumps and Grid Connected Solar Power Plants ," *NoticeInviti* 2019, [Online; accessed 28 July 2019]
45. Charging infrastructure of electrical vehicles Guidelines and Standards," 2019, [Online; accessed 28-July-2019] ngCommentsonGuidelines.pdf/, 2019, [Online; accessed 28 July 2019].
46. "Electrical vehicle charging station Guidelines by ministry of housing ," 2019, [Online; accessed 28 July 2019].
47. GS Li, XC Lu, JY Kim, KD Meinhardt, HJ Chang, NL Canfield, VL Sprenkle, "Advanced intermediate temperature sodium-nickel chloride batteries with ultrahigh energy densit", *Nat Commun* 2016;7:10683.
48. B. Zakerin, S. Syri, "Electrical energy storage systems: a comparative life cycle cost analysis", *Renew Sustain Energy Rev* 2015;42:569–96.
49. P. Keil, M. Englberger, A. Jossen, "Hybrid energy storage systems for electric vehicles: an experimental analysis of performance improvements at subzero temperatures", *IEEE Trans Veh Technol* 2016;65(3):998–1006.
50. J.B.Dunn, L. Gaines, J. Sullivan, M.Q. Wang, "The impact of recycling on cradle-to-Gate energy consumption and greenhouse gas emissions of automotive lithiumion batteries", *J Chem Educ, Environ Sci Technol Am Chem Soc Publ* 11 2012;46(22):12704–10.
51. L. Fang, H.Y. Luo, *Advanced DC/DC converters. Power electronics and applications series,* Singapore: CRC Press; 2003. p. 792.
52. Sulaiman N, Hannan MA, Mohamed A, Majlan EH, Daud WRW. A review on energy management system for fuel cell hybrid electric vehicle: issues and challenges. *Renew Sustain Energy Rev* 2015;52:802–14.
53. L. Li, J.B.Dunn, XX Zhang, L. Gaines , R.J.Chen, F. Wu, K. Amine,. "Recovery of metals from spent lithium-ion batteries with organic acids as leaching reagents and environmental assessment", *J Power Sources,* 7 2013;233:180–9.
54. D. Moon, J. Park, S. Choi, "New interleaved current-fed resonant converter with significantly reduced high current side output filter for EV and HEV applications", *IEEE Trans Power Electron* 2015;30(8):4264–71.

Index

Absorption, 11–13, 94, 224, 288, 299
Abundant, 82, 96, 100
Access, 18, 78, 82, 177, 211, 281, 313, 338
Accuracy, 40, 159, 163, 164, 168, 173, 175, 178
Actual-world, 34
Adequately, 14, 160, 299
Advancements, 72, 78, 133, 182, 214, 221, 320, 336
Affect, 37, 40, 50, 54, 59, 63, 159, 306, 327, 330
Affordable, 18, 94, 241
Agency, 2, 3, 19, 101
Algebraic, 181, 239
Algorithms, 52, 54, 64, 136, 175, 209
Anodes, 33, 53, 66
Anomalous, 59
Artificial, 8, 162, 166, 178, 186, 314
Auxiliary, 73, 140, 241

Balanced, 98, 233, 242, 329
Barriers, 22, 146, 157, 209, 222, 223, 235, 236
Battery-powered, 28, 319
Bio-carbon, 73, 80, 100
Biochar-based, 81–84, 87, 89, 91, 92, 95, 96, 98–102
Bioenergy, 102

Cadmium, 321
Calendar, 9, 41, 42, 67, 76
Calibration, 40

Capacitive, 89, 92, 100, 103, 133, 134
Catalysts, 141, 143, 155
Century-old, 297
Ceramic, 140, 144, 150
Charge-discharge, 8, 43, 73, 87, 92, 100, 171
Climate, 2, 20, 21, 27, 58, 102
Clipping, 218
Community, 69, 70, 229, 315
Conservation, 57, 94, 214, 218, 294
Copper, 45
Cumulative, 3, 226, 230–32, 287
Curtailment, 14, 216, 244, 248, 249, 252, 271, 278
Cyber-threats, 160

Data-driven, 162, 163, 165, 175, 177
Decades, 182, 216, 219, 319
Decarbonization, 93
Deregulation, 18, 182, 280, 283, 284, 301
Disaster, 52, 317, 318
Distillation, 148
Distinctive, 10, 92
Dual-energy, 181, 182, 192, 196

Earlier, 54, 139, 173, 217, 221
Eco-friendly, 73, 84, 87, 88, 317
Ecological, 160, 317, 318
Economical, 9, 138, 199, 203
Economic-emission, 210
Electrification, 21, 71, 133, 179
Electrocatalyst, 148, 150, 151

341

342 INDEX

Emergency, 3, 105, 106, 120, 122, 128, 215, 217, 218, 243, 286
Emission-free, 12
Environment, 12, 52, 71, 80, 94, 132, 134, 136, 216, 236, 237, 280, 297, 339
Error, 118, 159, 162, 163, 166, 173, 178
Estimated, 19, 162, 164, 165, 173, 284
Evolution, 27, 49, 50, 103, 214
Exposure, 17

Fabrication, 75
Facilitate, 15, 18, 20, 37–40, 56, 64, 78
Facing, 18, 146, 221
Flywheels, 75, 160, 289, 333, 334
Forecasted, 173, 225, 229, 230, 232–34
Friction, 292
Fuse, 51
Fuzzy, 106–108, 132, 134, 162, 163, 166, 177

Gasification, 81
Gasoline-fuelled, 286
Geographic, 330
Greenhouse, 2, 27, 160, 213, 220, 297, 336, 340
Grid-scale, 66, 177

Harvesters, 79
Heteroatom, 94, 100
Hierarchy, 302, 303
High-performance, 21, 81, 84, 87, 89, 92, 94, 99, 103, 104
Huge, 74, 90, 92, 162, 318, 328
Humidifier, 143
Hurdles, 73, 317
Hydrogen-oxygen, 136

Illustration, 30
Imminent, 297
Incentive, 187, 215, 242, 243, 249, 271, 281
Inertia, 8
Infiltration, 284

Instability, 10, 43, 139
Instant, 138, 200, 246
Interfaces, 16, 68, 79, 83, 319, 336
Intermittency, 2, 3, 15, 183, 189, 196, 213, 224, 235, 237, 287
Iron-chromium, 288
Irrigation, 211, 339
Isothermal, 6
Iteration, 303

Joint, 183, 209

Kilometers, 332
Kilowatts, 20

Laboratory, 156, 164, 315
Lagrangian, 168, 251, 252
Lead-acid, 13, 22, 24, 26, 76, 160, 211, 236, 241, 316, 333, 334
Life-cycles, 334
Lifespan, 52, 131
Lifetime, 24, 26, 31, 51, 65, 68, 76, 77, 100, 160, 289
Lithium-titanate-oxide, 33, 64
Load-following, 242
Load-shifting, 13, 219

Machine-learning-based, 66
Magnitude, 190, 250, 284
Matrix, 11, 81, 83, 101, 169, 170
Microgrid, 22, 73, 106–8, 110, 112, 131–33, 136, 156, 159–61, 175, 177, 181–92, 194–96, 198–200, 203, 205–12, 236, 315, 316
Micro-grid, 1, 3, 9, 16, 133, 208, 315
Micro-hydro, 133
Model-based, 162–66, 177, 178, 235
Morphologies, 84, 100
Motoring-generating, 203

Nanofibers, 12, 102
Nanomaterials, 73, 102, 104
Net-zero, 3, 287
Nickel, 53, 82, 299, 321

Nickel-cadmium, 13, 20, 26, 241, 288
Nickel–metal, 321

Obstacles, 99, 100
Off-grid, 3, 319
Optima, 303
Optimization-based, 285, 313
Oscillation, 119, 125, 128, 131
Outages, 211, 243, 296
Ownership, 324
Oxidation, 43, 49, 55, 79, 83, 96–98, 103, 113, 139, 144, 148

Panels, 329, 332
Paraffin, 11
Penetration, 15, 16, 18, 20, 161, 182, 220, 222, 284, 286, 300, 301, 314
Phase, 34–36, 92, 132, 289, 332
Phenomenon, 36, 38, 55, 61
Photovoltaic, 16, 177, 182, 209, 223, 283, 290, 339
Platform, 73, 82, 102, 119
Platinum, 144, 146, 148, 150, 151
Plugging, 298
p-phenylenediamine, 91
Precipitation, 45
Precision, 162, 164, 175
Price-based, 216, 294, 295
Propagation, 37, 71, 78
Properties, 13, 31, 33, 48, 62, 73, 81, 82, 94, 96, 99, 100, 148, 162, 175, 183, 281
Proportional, 75, 105, 115, 117
Pulse-charging, 34, 37

Quadratic, 168, 174–76, 189, 190, 244
Quantitative, 82
Quick-charging, 26, 33
Quite, 38, 148, 162, 168, 232–34, 330

Radiation, 230, 290
Rapid-charging, 57, 61

Reactants, 136, 144
Regulatory, 18, 222
Residues, 71, 84–86, 94, 99
Reversibility, 78, 83
Roundtrip, 4

Security, 1, 14, 18, 71, 72, 74, 93, 131, 223, 242, 337
Simulation, 18, 69, 70, 134, 188, 195, 304
Slack, 167, 168
Sodium, 75, 76, 91, 96, 241, 321
Sodium-sulphur, 13, 288
Steady, 32, 35
Stipulated, 293
Subsidies, 286, 317
Susceptible, 16, 48
Sustainability, 14, 103, 135, 149, 218, 313, 337
Swapping, 324, 339
Synthesized, 81, 92

Techno-economic, 22, 76, 77, 157, 211, 236, 283, 312, 316
Titanium, 33
Toxic, 98, 321
Transformers, 327
Transient, 112, 313
Transition, 2, 43, 45–47, 67, 78, 318
Tuned, 82, 110, 129, 131, 132

Ultra-capacitor, 9, 105, 115, 116, 136
Uncertainties, 18, 210, 211, 220–22, 229, 235
Utility-scale, 3, 19

Vacuum, 8
Volumetric, 27, 59, 91, 104

Wattage, 4
Watt-hours, 27

Also of Interest

Check out these other related titles from Scrivener Publishing

SMART GRIDS FOR SMART CITIES VOLUME 1, Edited by O.V. Gnana Swathika, K. Karthikeyan, and Sanjeevikumar Padmanaban, ISBN: 9781119872078. Written and edited by a team of experts in the field, this first volume in a two-volume set focuses on an interdisciplinary perspective on the financial, environmental, and other benefits of smart grid technologies and solutions for smart cities.

SMART GRIDS FOR SMART CITIES VOLUME 2: Real-Time Applications in Smart Cities, Edited by O.V. Gnana Swathika, K. Karthikeyan, and Sanjeevikumar Padmanaban, ISBN: 9781394215874. Written and edited by a team of experts in the field, this second volume in a two-volume set focuses on an interdisciplinary perspective on the financial, environmental, and other benefits of smart grid technologies and solutions for smart cities.

SMART GRIDS AND INTERNET OF THINGS, Edited by Sanjeevikumar Padmanaban, Jens Bo Holm-Nielsen, Rajesh Kumar Dhanaraj, Malathy Sathyamoorthy, and Balamurugan Balusamy, ISBN: 9781119812449. Written and edited by a team of international professionals, this groundbreaking new volume covers the latest technologies in automation, tracking, energy distribution and consumption of Internet of Things (IoT) devices with smart grids.

SMART GRIDS AND GREEN ENERGY SYSTEMS, Edited by A. Chitra, V. Indragandhi and W. Razia Sultana, ISBN: 9781119872030. Presenting the concepts and advances of smart grids within the context of "green" energy systems, this volume, written and edited by a global team of experts, goes into the practical applications that can be utilized across multiple disciplines and industries, for both the engineer and the student.

SMART GRIDS AND MICROGRIDS: Concepts and Applications, Edited by P. Prajof, S. Mohan Krishna, J. L. Febin Daya, Umashankar Subramaniam, and P. V. Brijesh, ISBN: 9781119760559. Written and edited by a team of experts in the field, this is the most comprehensive and up to date study of smart grids and microgrids for engineers, scientists, students, and other professionals.

MICROGRID TECHNOLOGIES, Edited by C. Sharmeela, P. Sivaraman, P. Sanjeevikumar, and Jens Bo Holm-Nielsen, ISBN 9781119710790. Covering the concepts and fundamentals of microgrid technologies, this volume, written and edited by a global team of experts, also goes into the practical applications that can be utilized across multiple industries, for both the engineer and the student.

INTEGRATION OF RENEWABLE ENERGY SOURCES WITH SMART GRIDS, Edited by A. Mahaboob Subahani, M. Kathiresh and G. R. Kanagachidambaresan, ISBN: 9781119750420. Provides comprehensive coverage of renewable energy and its integration with smart grid technologies.

Encyclopedia of Renewable Energy, by James G. Speight, ISBN 9781119363675. Written by a highly respected engineer and prolific author in the energy sector, this is the single most comprehensive, thorough, and up to date reference work on renewable energy.

Green Energy: Solar Energy, Photovoltaics, and Smart Cities, edited by Suman Lata Tripathi and Sanjeevikumar Padmanaban, ISBN 9781119760764. Covering the concepts and fundamentals of green energy, this volume, written and edited by a global team of experts, also goes into the practical applications that can be utilized across multiple industries, for both the engineer and the student.

Energy Storage, edited by Umakanta Sahoo, ISBN 9781119555513. Written and edited by a team of well-known and respected experts in the field, this new volume on energy storage presents the state-of-the-art developments and challenges in the field of renewable energy systems for sustainability and scalability for engineers, researchers, academicians, industry.

Energy Storage 2nd Edition, by Ralph Zito and Haleh Ardebili, ISBN 9781119083597. A revision of the groundbreaking study of methods for storing energy on a massive scale to be used in wind, solar, and other renewable energy systems.

Hybrid Renewable Energy Systems, edited by Umakanta Sahoo, ISBN 9781119555575. Edited and written by some of the world's top experts in renewable energy, this is the most comprehensive and in-depth volume on hybrid renewable energy systems available, a must-have for any engineer, scientist, or student.

Progress in Solar Energy Technology and Applications, edited by Umakanta Sahoo, ISBN 9781119555605. This first volume in the new groundbreaking series, Advances in Renewable Energy, covers the latest concepts, trends, techniques, processes, and materials in solar energy, focusing on the state-of-the-art for the field and written by a group of world-renowned experts.

A Polygeneration Process Concept for Hybrid Solar and Biomass Power Plants: Simulation, Modeling, and Optimization, by Umakanta Sahoo, ISBN 9781119536093. This is the most comprehensive and in-depth study of the theory and practical applications of a new and groundbreaking method for the energy industry to "go green" with renewable and alternative energy sources.

DESIGN AND DEVELOPMENT OF EFFICIENT ENERGY SYSTEMS, edited by Suman Lata Tripathi, Dushyant Kumar Singh, Sanjeevikumar Padmanaban, and P. Raja, ISBN 9781119761631. Covering the concepts and fundamentals of efficient energy systems, this volume, written and edited by a global team of experts, also goes into the practical applications that can be utilized across multiple industries, for both the engineer and the student.

INTELLIGENT RENEWABLE ENERGY SYSTEMS: *Integrating Artificial Intelligence Techniques and Optimization Algorithms,* edited by Neeraj Priyadarshi, Akash Kumar Bhoi, Sanjeevikumar Padmanaban, S. Balamurugan, and Jens Bo Holm-Nielsen, ISBN 9781119786276. This collection of papers on artificial intelligence and other methods for improving renewable energy systems, written by industry experts, is a reflection of the state of the art, a must-have for engineers, maintenance personnel, students, and anyone else wanting to stay abreast with current energy systems concepts and technology.

SMART CHARGING SOLUTIONS FOR HYBRID AND ELECTRIC VEHICLES, edited by Sulabh Sachan, Sanjeevikumar Padmanaban, and Sanchari Deb, ISBN 9781119768951. Written and edited by a team of experts in the field, this is the most comprehensive and up to date study of smart charging solutions for hybrid and electric vehicles for engineers, scientists, students, and other professionals.

Printed and bound by CPI Group (UK) Ltd, Croydon, CR0 4YY
05/09/2023
08109002-0001